International and Cultural Psychology

Series Editor

Anthony J. Marsella

For further volumes:
http://www.springer.com/series/6089

Douglas A. Vakoch
Editor

Altruism in Cross-Cultural Perspective

Springer

Editor
Douglas A. Vakoch
SETI Institute
Center for SETI Research
Mountain View, CA, USA

California Institute of Integral Studies
Department of Psychology
San Francisco, CA, USA

ISSN 1574-0455
ISBN 978-1-4614-6951-3 ISBN 978-1-4614-6952-0 (eBook)
DOI 10.1007/978-1-4614-6952-0
Springer New York Heidelberg Dordrecht London

Library of Congress Control Number: 2013937722

Printed on acid-free paper

Springer is part of Springer Science+Business Media (www.springer.com)

To Edna DeVore, for a career of altruistic leadership in science education

Foreword

Does altruism exist? Batson and Oleson (1991) identified Aristotle (384–322 B.C.E.), St. Thomas Aquinas (1227–1274), Thomas Hobbes (1588–1679), the Duke de la Rochefoucauld (1613–1680), Jeremy Bentham (1748–1832), Friedrich Nietzsche (1844–1900), and Sigmund Freud (1850–1939) as among those who attempted to answer that question. The consensus is that altruism is a real phenomenon.

The phenomenon of altruism was examined also by the Roman poet Lucretius (2007). Lucretius was a materialist (no supernaturals are needed) and thought that hedonism is a central value explaining human behavior. But he also thought that all humans form one family, and it is the obligation of all of us *to help* all others to reach hedonism.

The need to help others to be happy is also found in Triandis (2009b) who argued that the purpose of life is to help as many people as possible to be healthy (both physically and mentally) and happy, so they can live a long time without destroying the environment. One can argue, as many commentators on Lucretius have done, that helping others to be happy is an altruistic act, since it *benefits others and also imposes some cost on the actor.*

All definitions of altruism include the benefit received by another. However, not all definitions include the cost to the actor. For instance, Bernhard, Fischbacher, and Fehr (2006) see behavior that favors the in-group more than the out-group as "parochial altruism." Batson and Oleson (1991) also do not include the cost in their definition.

The authors of the present important book also include extensive discussions of the best definitions of altruism and empathy and, in some cases, extend the concepts to include cooperation, generosity, and spirituality. Lee et al. describe the person who is the most altruistic by using the metaphor of water. Water is beneficial without asking to be repaid; it settles at the lowest level, as the altruistic person is modest; it is flexible, yet it can be powerful; it is transparent just as the person is honest; it is gentle but can persevere.

An important theme in this book is that behaviors that are altruistic in one culture may not be altruistic in others. This raises the question: What are the most important ways in which cultures differ?

Contemporary cross-cultural psychology (Kitayama & Cohen, 2007; Markus & Kitayama, 1991; Triandis, 1989, 1994, 1995, 2009a, 2009b; Triandis & Gelfand, 2012) emphasizes the contrast between collectivist (gemeinschaft, sociocentric, communal relations, collaterality, community, communitarian, interdependent, traditional) and individualist (gesellschaft, self-emphasis, person-centered, agency, independent, modern) cultures. Hundreds of publications have used this contrast. However, three other dimensions have also proven important:

1. Cultural simplicity (as found among hunters and gatherers) vs. cultural complexity (as found in information societies) (Chick, 1997)
2. Cultural tightness vs. cultural looseness (Gelfand et al., 2011; Pelto, 1968; Triandis, 1994), that is, cultures with many norms imposed tightly (e.g., if one does not do what is expected, one is killed) vs. imposed loosely (e.g., if one does not do what is expected, people smile), for example, Taliban vs. rural Thailand
3. Vertical (highly hierarchical) vs. horizontal (less hierarchical) cultures

Triandis (1995) used four attributes to define collectivism. First, collectivists define the self as a member of a collective, such as family, village, or religion. If asked to complete sentences that start with "I am…" they complete them by mentioning a group or a relationship (e.g., I am a cousin). They are greatly concerned about the welfare of members of this collective. Second, collectivists give priority to the goals of their in-group rather than their personal goals. For instance, if offered a job that their family does not like, they reject the job. Next, the behavior of collectivists is determined by group norms to a greater extent than by personal attitudes. For instance, they are more likely to do their duty than to do what they like. Finally, when collectivists do not like their group, they tend to stay with it, rather than look for another group.

There are many kinds of collectivism and individualism, the most important being the vertical (V) vs. the horizontal (H) variety (Shavitt, Torelli, & Riemer, 2011; Triandis, 1995; Triandis & Gelfand, 1998).

The correlates of simplicity–complexity are primarily ecological (Triandis, 2009a). Especially important is the size of the population settlement. Simple cultures consist of bands of 50–200 individuals, while complex societies have millions of members. It is reasonable to assume that cognitive simplicity and cultural simplicity are related.

Gelfand et al. (2011) found that tightness is correlated with population density, scarcity of resources, terrorism, natural disasters, disease, great religiosity, autocracy, close monitoring of social behavior, many prohibitions, and censorship.

The different kinds of collectivism and individualism result in four kinds of societies: VC (vertical collectivist), HC (horizontal collectivist), VI (vertical individualist), and HI (horizontal individualist). The VC pattern is found in most traditional societies, such as rural China or India. The major value is conformity to the authorities. Bond and Smith (1996) found more conformity in collectivist cultures, as measured by the Asch paradigm, than in individualist cultures.

The HC pattern is found in the Israeli kibbutz. The major value is cooperation. The VI pattern is found in Western Europe and North America, as well as in academia and major corporations, where achievement and competition are the important values. The HI pattern is found in Scandinavia, Australia, and New Zealand, where the major value is the *uniqueness* of the individual; however, individuals do not want to "stick out" (Triandis & Gelfand, 1998, 2012). In fact, Australian psychologist Norman Feather argues that in Australia tall poppies are cut down.

Affluence and globalization tend to shift cultures from VC to HI. I offer the hypothesis that national altruism will be greater and more frequent in HI cultures and least common in VC cultures. In fact, it would be good to test if the hierarchy might be HI (Scandinavians), HC (kibbutz), VI (Western Europe, academia), VC (rural India). Data on the donors of foreign aid tend to support this hypothesis. The top seven donors are in Scandinavia, Luxemburg, Belgium, and Ireland (*World Almanac and Book of Facts*). Of course, in order to be a donor a country has to be wealthy, and individualism is correlated with affluence (Hofstede, 2001). There is also evidence (Gergen, Morse, & Gergen, 1980) that in most cultures the higher social classes are more altruistic than the lower classes, and this may simply reflect the difference in affluence. If one has ample resources, one can afford to be more altruistic. Gergen et al. also report that poor donors are more attractive than rich donors, which suggests that while rich donors help more frequently they are not necessarily the most attractive helpers.

Obviously, more research is needed to test the hypothesis presented above. If the hypothesis is supported, it may be that horizontality is more important than individualism as a factor that increases altruism. Perhaps people who spend too much time worrying about their position in a hierarchy do not have enough energy left to help others. It may also be the case that horizontals feel closer to others and can feel empathy more easily than verticals. All these hypotheses invite further research.

Triandis (1994, pp. 221–225) has reviewed studies that determined that there are cultural differences in the probability that a person will help others. For example, while in most cultures people are more likely to help an in-group than an out-group member, this difference is greater in collectivist than in individualist cultures. However, anomalies do emerge. For instance, in Greek villages, people might be more willing to help a stranger than a fellow villager. This finding may reflect the tendency to provisionally categorize foreigners as in-group members (Triandis, 1972). The status of foreigners is tested, and if those foreigners are helpful, they are permanently categorized as an in-group member, but if not helpful, they are categorized as an out-group member and will not be helped in the future.

The act of helping may be perceived as more dangerous in some contexts, such as where there is much crime, than in others. Thus, in cultures that are dangerous, there may be less helping. The implications associated with helping also differ with culture. For instance, the Japanese hesitate to help a stranger more than do Americans. The act of helping creates a relationship that may have further costs.

The person's obligation to help is also different in different cultures. In India, people often feel greater obligations about helping than in America. In Islam, helping is obligatory; thus, thanking a person for helping is insulting, since one does not thank a person for fulfilling a religious requirement (Chiu & Hong, 2006, p. 102).

Chiu and Hong (2006, pp. 177–178) review evidence about the relationship between culture and helping behavior. In North America, reciprocal helping is seen as a personal choice rather than an obligation. Thus, one is more likely to help a liked than a disliked other. In India, reciprocal helping is a moral obligation and not a personal choice. Thus, in India, the degree of liking is much less important as a factor determining the probability of helping. Also, Americans rate the target's liking to help as less important in the reciprocal helping than in the spontaneous helping condition.

The paragraphs that follow offer comments about the chapters in this book, with special emphasis on the way the data collection methods may have influenced the findings.

Sandi W. Smith, Maria Knight Lapinski, Mary J. Bresnahan, and Stacy L. Smith point out that behaviors that are considered altruistic in one culture may not be viewed in that way in other cultures—an important point. In fact, there is even some evidence that "helping" is not always considered desirable (Gergen, Morse, & Gergen, 1980).

M. Kathryn Coe and Craig T. Palmer suggest that over time, traditions become less important and other factors become more important. Since collectivist cultures are more traditional, that would suggest that over time culture may become a less important factor influencing altruism. They also present an important evolutionary perspective on altruism.

Judith L. Gibbons obtained unexpected results. Boys helped more than girls, and urban children helped more than rural children. The latter finding is inconsistent with the findings of a meta-analysis of 65 tests of the hypothesis which found that rural populations help more than urban populations (Steblay, 1987). One needs to explore if the methodology of using photographs taken by the children may be a factor in these findings. Different methods seem to provide different kinds of data: For instance, when Gibbons measured how frequently the children used "we," the rural children did so more often than the urban children. This emphasizes once more the importance of using multi-method approaches in data collection.

Alexandra Arkhipova and Artem Kozmin present a fascinating analysis of the presence of altruism in fairytales. They show that altruism is found in only about 3% of a large sample of fairytales. It would be interesting to know the frequency of aggression, submission, domination, avoidance, support, and other social behaviors in such tales. They also show the geographic distribution of particular themes in fairytales. For instance, saving the self rather than waiting for help from others appears to be linked to individualistic cultural patterns.

Henrietta Grönlund examines volunteering, which certainly is an altruistic act. The finding that egalitarian values are related to more volunteering fits the hypothesis that I presented earlier.

Yueh-Ting Lee, Wenting Chen, and Sydney Xinni Chan show how Daoism is related to altruism. The discussion of the intensity, extensity, duration, purity, and adequacy of the altruistic acts provides excellent avenues for more sophisticated research on altruism. The finding that Americans are more altruistic than Chinese seems to fit the hypothesis I presented earlier. The finding that females are more altruistic than males is inconsistent with the findings obtained in Guatemala by Gibbons, and again suggests that it is important to explore how the methods used may result in different kinds of findings.

Abhik Gupta presents a captivating exploration of how altruism is embedded in Indian religions, with special emphasis on *metta* (loving-kindness). I found the discussion of the role of Asoka, around 260 B.C.E., especially enlightening, since Asoka increased the culture's emphasis on altruism.

Sangeetha Menon also provides information about the way altruism operates in Indian philosophy and explores the evolutionary benefits of altruism.

Joan Koss-Chioino shows how spiritual transformation may result in healing people in distress. She finds altruism at the core of spirituality and provides a superb discussion of empathy.

Garry Chick examines cargo cultures which include the sponsorship of religious village celebrations. The *mayordomos* who do most of the work in such celebrations are altruistic. He also includes a discussion of the relevant neuroscience.

Lewis Aptekar draws from his fieldwork with street kids and examines various dilemmas that occur when humanitarian workers deal with difficult populations.

In short, this volume provides an excellent exploration of the way altruism is similar and different across cultures and offers a rich variety of perspectives from different cultures, presents new concepts and methods for the study of altruism, and opens new avenues for research on the relationship between culture and altruism.

Champaign-Urbana, USA Harry C. Triandis

References

Batson, C. D., & Oleson, K. C. (1991). Altruism. In *Encyclopedia of human biology, 1,* 197–207.

Bernhard, H., Fischbacher, U., & Fehr, E. (2005) Parochial altruism in humans. *Nature Letters, 442,* 912–915.

Bond, R., & Smith, P. B. (1996).Culture and conformity: A meta-analysis of studies using Asch's (1952b, 1956) line judgment task. *Psychological Bulletin, 119,* 111–137.

Chick, G. (1997). Cultural complexity: The concept and its measurement. *Cross-Cultural Research, 31,* 275–307.

Chiu, C-Y., & Hong, Y-Y. (2006). *Social psychology of culture.* New York: Psychology Press.

Gelfand, M. J., et al. (2011). Differences between tight and loose cultures: A 33-nation study. *Science, 332,* 1100–1104.

Gergen, K. J., Morse, S. J., & Gergen, M. M. (1980). Behavior exchange in cross-cultural perspective. In H. C. Triandis, & R.W. Brisliin (Eds.), *Handbook of cross-cultural psychology* (Vol. 5, pp. 121–153). Boston: Ally and Bacon.

Hofstede, G. (2001). *Culture's consequences* (2nd ed.). Thousand Oaks, CA: Sage.

Kitayama, S., & Cohen, D. (2007). *Handbook of cultural psychology.* New York: Guilford Press.

Lucretius (2007). *De Rerum Natura* (The nature of things) (A. E. Stallings & R. Jenkins, Trans.). New York: Penguin Books.

Markus, H., & Kitayama, S. (1991). Culture and self: Implications for cognition, emotion, and motivation. *Psychological Review, 98,* 225–253.

Pelto, P. J. (1968). The difference between "tight" and "loose" societies. *Transaction,* 37–40.

Shavitt, S., Torelli, C. J., & Riemer, H. (2011). Horizontal and vertical individualism and collectivism: Implications for understanding psychological processes. In M.J. Gelfand, C. Chiu, & Y. Hong (Eds.), *Advances in culture and psychology* (Vol. 1). New York: Oxford University Press.
Steblay, N. M. (1987). Helping behavior in rural and urban environments: A meta-analysis. *Psychological Bulletin, 102*, 346–358.
Triandis, H. C. (1972). *The analysis of subjective culture*. New York: Wiley.
Triandis, H. C. (1989). Self and social behavior in differing cultural contexts. *Psychological Review, 96*, 506–520.
Triandis, H. C. (1994). *Culture and social behavior*. New York: McGraw-Hill.
Triandis, H. C. (1995). *Individualism and collectivism*. Boulder, CO: Westview Press.
Triandis, H.C. (2009a). Ecological determinants of cultural variations. In R. W. Wyer, C-Y. Chiu, & Y-Y. Hong (Eds.), *Understanding culture: Theory, research and applications* (pp. 189–210). New York: Psychology Press.
Triandis, H. C. (2009b). *Fooling ourselves: Self-deception in politics, religion, and terrorism*. Westport, CT: Praeger.
Triandis, H.C., & Gelfand, M.J. (1998). Converging measurement of horizontal and vertical individualism and collectivism. *Journal of Personality and Social Psychology, 74*, 118–128.
Triandis, H.C., & Gelfand, M. (2012). A theory of individualism and collectivism. In P.A. M. Van Lange, A. W. Kruglanski, & E. T. Higgins (Eds.), *Handbook of social psychological theories* (Vol. 2, pp. 498–520). Los Angeles, CA: Sage.

Acknowledgments

To the contributors of chapters that appear in this volume, I especially appreciate the depth and innovation of the work they share here. These authors deserve special thanks for engaging one another in dialogue during the process of writing their chapters, as reflected in the numerous cross-references throughout the book.

Over the past 15 years, many colleagues from the SETI Institute have shared with me their insights into the universality of altruism. I especially thank Molly Bentley, Anu Bhagat, John Billingham, Edna DeVore, Frank Drake, Sophie Essen, Andrew Fraknoi, John Gertz, Gerry Harp, Jane Jordan, Ly Ly, Dave Morrison, Chris Munson, Chris Neller, Tom Pierson, Karen Randall, Jon Richards, Pierre Schwob, Seth Shostak, and Jill Tarter.

More recently, I warmly acknowledge the administration, faculty, staff, and students of the California Institute of Integral Studies (CIIS), especially for support from Joseph Subbiondo, Judie Wexler, and Bryant Welch. The work of editing this volume was made possible through a generous sabbatical leave from my other academic responsibilities at CIIS in the spring of 2012. In addition, I thank Harry and Joyce Letaw, as well as Jamie Baswell, for their support of this work.

For help in launching this project, I thank Rob Veneziano, who chaired the 34th Annual Conference of the Society for Cross-Cultural Research. By including the symposium "Altruism in Cross-Cultural Perspective" as part of that conference, Rob fostered a meeting where several of the chapters in this book were first presented in nascent form. I especially appreciate Rob's patience as we added more and more contributors to this dialogue between cross-cultural researchers and specialists in altruism, ultimately leading to multiple, interrelated sessions.

I am grateful to Lou Castrovinci for so capably copyediting each chapter, enhancing the consistency across chapters while still encouraging authors to keep their own voices. Without Lou's organization, care, and insight, this book would have taken much longer to complete.

For shepherding the book through the editorial process at Springer I am indebted to Sharon Panulla. Also at Springer, Sylvana Ruggirello has my gratitude for helping to move the book swiftly and efficiently through the review process and into production. Ejaz Ahmad of SPi Global has my appreciation for faithfully overseeing all aspects of copyediting, layout, and indexing.

I thank Anthony J. Marsella for including this volume in the Springer series International and Cultural Psychology, for which he serves as general editor.

Finally and most importantly, to my wife Julie Bayless, I am grateful in more ways that I can or will share here. Thank you, forever.

Mountain View, CA, USA Douglas Vakoch

Contents

Chapter 1
Altruism in Its Personal, Social, and Cultural Contexts: An Introduction

Juris G. Draguns

1.1 The Nature of Altruism: Introductory Considerations

1.1.1 What Is the Fundamental Source of Altruism?

Human beings are capable of enhancing the well-being of other persons, sometimes heroically, even at the cost of their own lives. Why they engage in altruistic acts, from the mundane to the exceptional, is one of the great unanswered questions about the human condition. On the basis of a thorough philosophical inquiry, Penner (1995) boldly asserted that "altruism is an innate capacity to be tapped, rather than something that must be farmed out of the rocky soil of egoism. Indeed, the great religions in different ways affirm that agape, karuna, jen – empathetic care for beings – is not only a possible human quality but one that is mysteriously constitutive of the structure of being itself, a quality that dwells at the very heart of reality" (p. 113). Penner's view, however, remains controversial. Over centuries, it has been held only by a minority of philosophers and psychologists. Indeed, the implicit assumptions of psychological science and its fundamental tenets, such as the principle of reinforcement, posit the primacy of the opposing position of universal egotism. Without endorsing it, Batson (1991, p. 2) describes its basic tenet as follows: "Everything we do, no matter how noble and beneficial to others, is really directed toward the ultimate goal of self-benefit." In this formulation, altruism is a derivative of egotism.

Somewhat paradoxically, questions about the primacy of egotism over altruism arose when altruism became object of empirical social science. Piliavin and Charng (1990, p. 61) discerned "a 'paradigm shift' away from the earlier position that behavior that appears to be altruistic must, under scrutiny, be revealed as reflecting

J.G. Draguns (✉)
Pennsylvania State University, Psychology Department, University Park, PA 16802-1515, USA
e-mail: jgd1@psu.edu

D.A. Vakoch (ed.), *Altruism in Cross-Cultural Perspective*,
International and Cultural Psychology, DOI 10.1007/978-1-4614-6952-0_1,

egotistic motives. Rather, theory and data now are more compatible with the view that true altruism – acting with the goal of benefiting another—does exist and is part of human nature."

Over the ensuing three decades, Piliavin and Charng's views have received further support from the results of a systematic research program of more than 30 experimental studies in social psychology in which altruism-based hypotheses were upheld over predictions based on egotism (Batson, 2010). On a more general plane, Batson (2010, p. 24) concluded that "we humans are more social than we have thought. Other people can be more to us than sources of information, gratification, and reward as we seek our own welfare. We have the potential to care about them for their sakes, not simply for our own." Thus, altruism plays a pivotal part in the human condition, and we must now address the complexities and ambiguities of altruism as a concept.

1.1.2 Altruism: Its Scope and Gist

Batson (2011) defined altruism as "a motivational state with the ultimate goal of increasing another's welfare." Note that in this definition altruism is defined in terms of motivation and not behavior. Yet, in general usage, it encompasses both. Altruistic intentions do not qualify unless they are consummated in action nor can action be considered altruistic if it is incidental, unintended, or unmotivated. Unconscious altruism is an oxymoron as is passive or inactive altruism.

Moreover, in both empirical social science research and philosophical discourse, emphasis has been placed on the most risky and dramatic instances of altruism. Harboring a Jewish family during Nazi occupation in World War II, rescuing a child from a burning building, or putting one's career and family life on hold in order to care for the sick in Africa are some examples that come to mind. However, Bierhoff (2001) describes volunteerism as a planned and organized expression of altruism. In this volume, altruism is investigated in organizing and conducting community rituals in Mexico (Chap. 11 by Chick), engaging in socially beneficial volunteering at various sites in the world (Chap. 6 by Grönlund), and being spontaneously helpful in one's home and neighborhood beyond one's chores and duties in the Mayan communities of Guatemala (Chap. 4 by Gibbons). Altruism encompasses both heroic, self-sacrificial acts and relatively inconspicuous and seemingly mundane instances of intentional helping, provided they are undertaken without the prospect of direct and indirect reward. Activation of altruism, as Koss-Chioino demonstrates in Chap. 10, serves as a major avenue of spiritual transformation in preparation for traditional healing in Mexico. In Chap. 2, Smith, Lapinski, Bresnahan, and Smith append an additional specification: a person's action should be regarded as altruistic only if it is intended to benefit others beyond the general requirements of helpfulness and benevolence that are associated with one's role within the society. This criterion is subtle and difficult to pin down in specific instances. Thus, the threshold for altruism may vary with the individualistic versus collectivistic nature of the society in question.

1.2 Altruism as an Object of Social Science Investigation

1.2.1 Issues, Topics, Methods, and Findings

The expansion of humanistic and positive psychology has powerfully stimulated conceptual and research interest in altruism as a component of the human potential for constructive, helpful, and benevolent action. A number of areas of investigation have emerged and are being vigorously pursued. We will introduce them briefly below.

1.2.1.1 Empathy

Empathy has been described as the vicarious experience of another person's emotional state (Staub, 2003) which is often accompanied by adapting that person's perspective in appraising his or her self, situation, and world (Draguns, 2007). Being able to tune in to another human being's affective and cognitive state may be a potent facilitator of altruistic motivation and helpful action, especially if that person experiences distress (McCauley & Bock, 2004; Royzman & Kumar, 2001). Several authors hypothesized and investigated the relationship between empathy and altruism, for the most part with positive results (Eisenberg & Miller, 1987; Krebs, 1975; Staub, 2003). Batson (1991, 2011) explicitly formulated and systematically tested the empathy-altruism hypothesis, which states that empathic concern produces altruistic motivation. For empathic concern to be experienced, two conditions must be met: the other person must be perceived as being in need, and that person's welfare must be valued. As Batson (2011, p. 29) notes, "The more empathy felt for the person in need, the more motivation to have the need removed." Over close to three decades, Batson and his team of coworkers systematically tested the empathy-altruism hypothesis against formulations based on the competing position of egoism, "a motivational state with the ultimate goal of increasing one's own welfare" (Batson, p. 20). In more than 30 experiments, Batson and his colleagues confronted participants with a believable, though contrived, need situation and a real opportunity for helpful intervention, as exemplified by taking electric shocks instead of a person who finds the shocks exceedingly uncomfortable or helping a woman take care of a young child after his and her parents lost their lives in an automobile accident. The experimenters then manipulated both the level of empathic concern for the person in need and a variable that changes depending on whether the most effective means of helping is to reach either the altruistic goal of removing the suffering person's need or one of the competing egoistic goals (Batson, 2010). As Batson (2010) stated, "overall, results of these experiments have consistently turned out as predicted by the empathy-altruism hypothesis; results have failed to support any of the egoistic alternatives" (p. 23).

Agreement with this conclusion, however, is not unanimous. Other experimenters (Cialdini, Brown, Lewis, Luce, & Neuberg, 1997; Maner et al., 2002) interpreted their results as indicative of the operation of internalized rewards and fusion of self and other as the mainsprings of altruism. Nonetheless, Piliavin and Charng's (1990) earlier conclusion that altruistic motivation is not a derivative of more basic egotistical motives has been substantially strengthened. The question posed by Hoffman (1981, p. 121)—"Is altruism part of human nature?"—can be answered, somewhat cautiously, with a "yes." Moreover, Batson (2010, 2011) has also concluded that the empathic concern hypothesis overshadows the competing prosocial explanations, specifically those based on the collectivistic motivation to benefit one's group as a whole or on the desire to uphold the abstract principle that helping a person in distress is intrinsically good. Smith et al. in Chap. 2 arrive at a compatible conclusion: empathy is a pancultural component of altruistic motivation and helpful action.

Recent findings fall into place as additional extensions and validation and extensions of the empathy-altruism hypothesis, such as the greater readiness to help by persons of limited resources and lower social status (Kraus, Piff, Mendoza-Denton, Rheinschmidt, & Keltner, 2012) as well as by those who have experienced suffering (Vollhardt & Staub, 2011). Through childhood and adolescence, altruistic orientation increases with age, but its growth is especially promoted by the development of perspective taking, vicariously experienced emotions, and concern for others, all of which are aspects of empathy (Eisenberg, Carlo, Murphy, & van Court, 1995).

However, we can ask if empathy is a necessary condition for altruism to occur. The originators of the empathy-altruism hypothesis (e.g., Batson, 2010, 2011) do not consider it the only antecedent of altruism, just a prominent one. As such, it merits continued intensive investigation. Empathy is more readily experienced toward persons who are similar to the empathizer (Draguns, 2007). Help to strangers is less readily extended than to one's acquaintances, neighbors, friends, and family members. Yet there are numerous instances where both the agents and the beneficiaries of the altruistic transaction remain anonymous and mutually unknown. To account for these situations, the concept of empathy must be broadened. In this book, Gupta in Chap. 8 proceeds from Indian religious tradition in expanding the concept of empathy not only to all living beings, animal, and plant but even to inanimate nature. In the process, the very nature of empathy is transmuted from an exchange of thoughts and feelings within a dyad to imbuing virtually the entire natural world with human qualities. Along somewhat similar lines, Menon in Chap. 9 evokes India's Buddhist heritage to obliterate the distinction between self and other by overcoming desire and bringing about fusion of external and internal reality. In Chap. 7, Lee, Chen, and Chan introduce the tenets of "wateristic personality," rooted in Daoist philosophy; they developed a measure of altruism based on these conceptions and applied it to compare Chinese and Americans. Note, however, that in this formulation, empathy is not a major antecedent, let alone a determinant of altruism. Ultimately, however, the capacity to feel other humans' distress and to act helpfully in order to remedy it may stem from what Monroe (1996) called the perception of common humanity that enables some persons to transcend barriers based on culture, language, appearance, belief, and a host of other physical and psychological characteristics.

1.2.1.2 Altruism Across the Life Span

How does altruism develop in the course of the human life span? In Greece, Kavakoulis (1998) detected the earliest stirrings of prosocial behavior within the first year of life. This study, however, was based on parents' reports, and the findings need to be corroborated by other methods. Svetlova, Nichols, and Brownell (2010) relied upon more direct observational data as they traced toddler's prosocial behavior from instrumental to altruistic helping; findings point to the first indicators of prosocial action around the age of two, thereby corroborating the earlier observations by Zahn-Wexler, Radke-Yarrow, Wagner, and Chapman (1992). The results of longitudinal and cross-sectional studies reviewed by Bierhoff (2001) document a progression through childhood and adolescence toward an empathic orientation consistent with Kohlberg's stages of moral development and indicative of prosocial and altruistic motivation. More recently, Eisenberg, Cumberland, Guthrie, Murphy, and Shepard (2005) found that prosocial responses in adolescents were related to both perspective taking and age, a finding that is both elaborated and integrated in Chap. 4 by Gibbons. Moreover, it is plausible to expect that the parent-child interaction during the socialization process would affect the level of altruism in adults. This possibility was explored by Mikulincer and Shaver (2005), who reported that parents' secure attachment style predicted compassion and altruism in their children. More broadly, the warmth dimension (Rohner, 2004) arising from the experience of parental acceptance versus rejection in childhood is likely to be an antecedent of altruism and prosocial conduct in adulthood. In Chap. 3, Coe and Palmer conclude that altruism within the ingroup is deeply embedded in the unwritten and implicit codes of the traditional cultures based on kinship. These codes are transmitted through socialization.

1.2.1.3 Naturalistic Studies

Two major sociologists are associated with originating the investigation of altruism. Auguste Comte, the founder of the discipline, coined the term and contrasted it with egoism (Batson, 1991), and Pitirim Sorokin (1950), a prominent theorist and researcher in Russia and the United States, is credited with launching systematic empirical study of altruistic persons. Sorokin employed the extreme group methodology which has been destined to play a prominent role in altruism research ever since. Specifically, he set out to investigate two populations. One of them consisted of guests on a radio program that featured good neighbors, nominated and selected on the basis of their consistent and exceptional helpfulness and friendliness. In an innovative but arguably more controversial manner, Sorokin also turned to an unusual source of data: the lives of saints canonized by the Roman Catholic and Russian Orthodox churches, many of whom had performed prodigious feats of altruism in serving their fellow humans. In an unknown number of cases, however, their canonization may have been based on saintly qualities distinct from altruism such as spirituality, religious zeal, asceticism, and propagation of the faith. Sorokin's approach has been widely emulated. It has been exemplified by Post's (2003) by compilation of biographical vignettes of outstanding altruists, including Mahatma Gandhi, Martin Luther King Jr., Albert Schweitzer, and the Dalai Lama. In their large-scale multimethod study, S.P. Oliner and P.M. Oliner (1988) combined qualitative and quantitative data on rescuers of Jews during the Holocaust in several European countries. The inhabitants of the Protestant village of Le Chambon in France where hundreds of Jewish refugees found shelter during World War II (Hallie, 1979) were also studied and described by means of qualitative methods. Similar approaches were applied in research on known rescuers or "heroic helpers" (Staub, 2003, p. 291) in Nazi Germany and Poland, by London (1970) and Tec (1986), respectively.

S.P. Oliner and P.M. Oliner (1988) found that rescuers were more empathetic toward other persons' suffering and distress than their counterparts drawn from the population at large. They exhibited a greater sense of autonomy as well as a stronger sense of personal and social responsibility. Their childhood recollections emphasized the primacy of reasoned explanation over physical punishment. Other salient features of rescuers' socialization included modeling their parents' behavior and personal and emotional closeness to them. A major differentiating characteristic that emerged from the Oliners' research was extensivity, or the degree to which people are capable of empathizing with individuals outside of their ethnic, cultural, or national group. Several authors (Becker & Eagly, 2004; Bierhoff, 2001; Gilbert, 2003; Monroe, 1996, 2002; Oliner, 2002) singled out the sense of common humanity as the mainspring that enabled ordinary people to commit extraordinary feats of heroism. Reiterating these conclusions and expanding on them, Hunt (1990) stated that "circumstances made far less difference than the laboratory research would seem to suggest" (p. 202). It is the personal attributes that proved crucial; they included, as recapitulated by Hunt (1990), general positive mood, empathy, emotional expressiveness, prosocial orientation, and strong internal moral standards. An overlapping catalogue of traits has emerged from the study of a very different group of altruistic helpers, those who provided first aid to injured victims in traffic accidents in Germany (Bierhoff, Klein, & Kramp, 1991). Their characteristics included social responsibility and empathy as well as an internal locus of control.

Becker and Eagly (2004) further expanded this domain of research by focusing their attention, in addition to the rescuers of potential victims of persecution, to Carnegie Hero Medal recipients, recognized for voluntarily risking their lives while attempting to save the life of another. They also encompassed in their study that pose a significant, though markedly lower, risk for survival and well-being: kidney donors, Peace Corps volunteers, and Doctors of the World. Yet, all of these categories constitute examples of what one may call risky altruism. Becker and Eagly (2004) were especially interested in the gender difference in performing heroic or risky altruistic acts. They found that there were more men than women among the Carnegie medalists, but that women were more numerous among Holocaust rescuers, kidney donors, Peace Corps volunteers, and Doctors of the World. Intensive and extensive study of these and other similar populations is well worth continuing.

1.2.1.4 Experimental Studies

In addition to the naturalistic studies of altruism in extreme situations, as presented in the above section, social psychologists have become increasingly adept at recreating experimentally realistic situations that provide opportunities for engaging in helpful behavior or avoiding such engagement. The research program directed by Batson (2011) was described earlier in this Introduction as were its results. Studies focused on bystander intervention, experimentally produced or real life, have been an important avenue of information on the development of altruism (Bar-Tal, 1982; Eisenberg & Strayer, 1987) and in elucidating the complexities of the relationship between empathy and altruism (Claar, Boehnke, & Silbereisen, 1984; Eisenberg & Miller, 1987; Stotland, Mathews, Sherman, Hanson, & Richardson, 1978). The results of this research have generally strengthened the case for the robustness of altruism as a major source of human conduct that is not easily explained as a derivative of other, more basic motives.

1.2.1.5 Psychometric Methods

Between the two extremes of naturalistic and experimental study, researchers have addressed individual differences in altruism by devising paper-and-pencil as well as pictorial measures for its study (Boehnke, 1988; Boehnke, Silbereisen, Eisenberg, Reykowski, & Pulmonari, 1989; Eisenberg, Boehnke, Schuler, & Silbereisen, 1985). Risky altruism tends to occur at unique conjunctions of space and time; it is an exceptional act that is inherently difficult to foresee. It can rarely be investigated other than retrospectively, and it defies comparison. For very different reasons, the intricate and fragile reconstructions and improvisations on which many experiments pertaining to altruism are based are difficult to replicate across national and cultural barriers. This is not the case with the Prosocial Motivation Questionnaire (PMQ) developed by Boehnke et al. (1989) that has been successfully applied in several languages and countries of Europe. The same considerations apply to the Questionnaire of Emotional Empathy (Mehrabian & Epstein, 1972), one of the first validated self-report measures of this construct. The flexibility and adaptability of psychometric methods argues for their continued use in research on altruism.

An important step has been taken by Lee, Chen, and Chan in Chap. 7 of this volume. In contrast to the generic European notion of altruism on which PMQ is based, Lee et al. proceed from Daoist wateristic philosophy, which includes altruism as one of its central features, in constructing their scale. To be sure, such measures are not destined to become major or exclusive tools for the assessment of individuals within and across all cultures. However, they do provide a distinct and valuable perspective on altruism as it is experienced and cognized across persons in the culture in which these constructs have originated.

1.2.1.6 Heritability

In an early review, Rushton (1984) summed up the evidence from experimental, naturalistic, and self-report studies to conclude that altruism constituted a human personality trait. Some people, Rushton maintained, are consistently generous, kind, and helpful in their dealings with other persons. Moreover, in a comparison of a large number of twin pairs, Rushton, Fulker, Neal, Nias, and Eysenck (1986) found a substantial hereditary component in individual differences in altruism and empathy as measured by questionnaires. More recently, Knafo and Israel (2010) examined seven studies of children and seven of adults, all of which except one showed evidence for the heritability of prosocial behavior. These findings are compatible with an upsurge of interest in the biological substrate of altruism (Hein & Singer, 2010) that is represented in Chap. 4 by Gibbons and Chap. 11 by Chick. Current research is focused on the hormone oxytocin that has been found to play a role in the activation of a gamut of prosocial behaviors, including altruism and empathy. Additional evidence comes from recent neuroimaging studies that have identified brain regions involved in arousal and experience of altruism. These findings have opened a new frontier for investigating the genetic potential for altruism. It would be hasty to say that human beings are programmed at birth for altruistic motives, feelings, and acts which only await environmental triggers to be activated. More likely, there are substances and circuits, some of which are situationally aroused and others which are, to varying degrees, part of the person's genetic makeup. They may interact with altruism in a variety of ways and may account for an as yet unknown share of individual differences in altruism.

1.2.1.7 Culture

Systematic investigation of the relationship between altruism and culture is still in its infancy, even though expressions of altruism, as all complex social acts, are embedded in their contexts. As already indicated, altruism is fostered, channeled, or impeded by the socialization experience, and socialization varies across culture. Values are a powerful influence on expressions of altruism (Oliner, 2002; Schwartz, 2010) and they are shaped by culture. Whether the recipient is a member of the ingroup greatly matters in extending help and support, but does so to different degrees across cultures. It appears highly unlikely that there are cultures whose members are uniformly and consistently helpful and altruistic, even when altruistic acts involve danger. Equally implausible is the existence of cultures in which altruism is never encountered or practiced. Between these two extremes are situated the existing cultures of the world, wherein the incidence of altruism is codetermined by the characteristics of the person who is performing the altruistic act, those of the beneficiary of this action, and the multiple features of the situation in which the act occurs. In Chap. 2, Smith et al. propose a more elaborate scheme. It features three components that are culturally variable: whether the actor would be inclined to blame oneself or be socially blamed if he or she did not engage in this act, and the extent to which it would benefit the recipient. Two additional components are posited as culturally invariant: empathy with the recipient and cost to the actor. This model may serve as a workable general guide for the systematic investigation of altruism in culture.

1.2.2 Directions for Future Research: What We Do Not Know Yet

Over the past several decades, substantial advances have been made in the investigation of altruism, both in the culturally historic arena and in the laboratory setting. Empirically based, answers, as yet for the most part tentative, have begun to emerge. Below is a list of current research areas some of which are rigorously pursued; others await more intensive investigation.

1.2.2.1 Consistency

How consistent are the expressions of altruism across contexts and throughout the life span? An early data-based answer to this question by Rushton (1984) emphasized the traitlike nature of altruism, marked by consistency in generosity, kindness, and helpfulness. This catalogue of characteristics greatly overlaps the traits of rescuers identified by Oliner and Oliner (1988) who, moreover, concluded that circumstances made a lot less difference than dispositions. However, almost three decades have elapsed since these conclusions were reached, and the field is ripe for another phase of investigation, perhaps eventually by multivariate techniques and culminating in the meta-analysis of accumulated data. Intuitively, a wide gulf separates studies of bystander intervention from rescuing other human beings at the peril of death. Therefore, a monofactorial solution to the problem appears exceedingly unlikely. Researchers' attention should be concentrated on identifying the various complex person-by-situation interactions that may emerge in such analyses.

1.2.2.2 Unconscious Motives

If pluralism of altruistic motives underlying helpful action emerges as the provisional explanation of the available findings, attention should be expanded to the hidden, camouflaged, and paradoxical sources of altruism. There is suggestive evidence, recapitulated earlier in this introduction, that altruistic individuals enjoy close personal relationships to others that are mutual and rewarding. However, practicing psychotherapists (e.g., Hunt, 1990) have noted that altruism is sometimes embedded in personalities characterized by loneliness and lack of fulfillment. The compensatory or defensive nature of altruism has so far been rarely mentioned or explored in social psychologists' writings on altruism, nor has it been much studied clinically, even at the case level. Nonetheless, it is worth noting that Vaillant (1977) in his compendium of defense mechanisms included altruism, albeit among the mature mechanisms of defense. Other varieties of such defenses are sublimation and humor. In Vaillant's view, altruism is a disposition that may in some cases result from the person's attempts to reduce the impact of difficult and stressful personal dilemmas. Vaillant, however, did not imply that altruism is invariably traceable to these hidden unconscious roots. At this point, psychodynamic sources of altruism remain a little-studied facet of the relationship between the person, his or her multiple motives, and the prosocial actions that spring from them. Recently, Tolmacz (2010) has made a start by identifying four varieties of concern and tracing them to drive and relational models of psychoanalysis.

1.2.2.3 Intrapersonal Conflict

The role of intrapersonal conflict in prosocial behavior, especially when such action is heroic or risky, has not yet received the emphasis that it deserves. Such conflict is frequent and, in extreme cases, inescapable. Whether or not to give to charity and how much, or whether or not to volunteer one's time and effort to a community project evokes a conflict that is relatively easy to resolve. This is not the case with the more consequential, sometimes literally life and death, conflicts, as between avoiding risk or harm to oneself and abandoning other human beings to certain or probable torture, mistreatment, or death or, conversely, between saving others and risking one's life in the process. Accounts of the experience and resolution of these approach- avoidance and double approach-avoidance conflicts would be unique contributions to the conflict-resolution and decision-making literature and would also shed light on the ultimate sources of prosocial behavior under the most trying circumstances.

1.2.2.4 Personality Traits

Little work so far has been done to relate altruism to the basic, empirically founded dimensions of personality, such as the Big Five factors that have been systematically investigated across cultures by McCrea and Allik (2002). Two of the five would appear to be especially relevant: agreeableness, characterized by ability and willingness to trust others and to engage in friendly and cooperative interaction with them, and conscientiousness, with its emphasis on moral obligations. In Israel, Sommerfeld (2010) included Big Five measures and other personality scales in her research on generosity and reported a complex array of findings, generally in keeping with the above expectations. Additionally, she found a negative correlation between spontaneous generosity and most aspects of neuroticism, the Big Five dimension that principally taps a person's vulnerability to internal and external stress.

1.2.2.5 Cultural Dimensions

Similarly to personality traits, cultural dimensions, such as those intensively investigated throughout the world by Hofstede (2001), should also be considered in relation to altruism. As yet, little research along these lines has been done, and the index of Hofstede's *magnum opus* does not contain an entry for altruism. Of Hofstede's five dimensions, femininity is the most likely to be linked with prosocial variables. A caring attitude is fostered in feminine cultures, in contrast to the competition and achievement prized in masculine cultures, thereby favoring altruism. In regard to individualism-collectivism, the most thoroughly investigated Hofstede's dimension, its relationship with altruism may prove to be interactive and complex and may have less to do with differences in the absolute levels of altruism than with the contexts of its manifestation. This expectation is consonant with the thrust of Chap. 2 by Smith et al., who draw a sharp line between the manifestations of altruism in individualistic cultures, where helpful acts stem from personal decision and choice, and those in collectivistic cultures, in which they are performed as a matter of role and obligation. Similarly, Schwartz (2010) has asserted that culturally characteristic basic values shape prosocial behavior by influencing the direction of motivation and by establishing priorities among values. Human Relations Area Files (HRAF) contain potentially valuable information on expressions of altruism in its various cultural contexts. As yet, HRAF have rarely been tapped for this purpose. For a start, one may ask whether there are ethnographies on traditional cultures that are at the opposite poles of helpfulness: extremely high in altruism and virtually bereft of it. The contrast drawn by Palmer and Coe in Chap. 3 between kinship-based societies and the more modern cultures governed by explicit laws and precepts may be investigated by means of coded HRAF samples. In addition to archival sources of information, there is a novel and as yet little known method of quickly collecting virtually worldwide data by a network of international collaborators that can be applied to the cross-cultural investigation of altruism and its concomitants and correlates. Electronic communication has enabled teams of researchers to gather, pool, and compare information simultaneously on such topics as preferences in selecting marriage partners (Buss et al., 1990), social axioms (Bond et al., 2004), and romantic attachments (Schmitt et al., 2004) in samples from 37, 41, and 62 countries and regions, respectively. The studies reported are based on the availability of collaborators; necessarily, their participants constitute samples of convenience. This innovative method is more applicable at the early rather than late stages of investigating a topic, unless the problems of representativeness, within samples and across countries, are resolved. National indicators of humanitarian aid by governments and voluntary nongovernmental organizations (NGOs) can also be utilized as rough approximations of national levels of altruism, provided they are adjusted for gross national product and are based on democratic appropriation and/or voluntary collection procedures. And in an abrupt shift of gears, innovative researchers may turn to the folkloric heritage of world's cultures. In Chap. 5, Arkhipova and Kozmin explore folktales, some of them of ancient origin, in order to trace possible connections between altruism in its contemporary modes of expression and their remote roots in antiquity.

1.2.2.6 Cross-Cultural Research

Explicitly designed cross-cultural research relevant to helpfulness and altruism has so far been rarely reported. As examples of such studies, we can point to a tricultural study of willingness to mail a stamped letter in Athens, Boston, and Paris (Feldman, 1968); a comparison of prosocial moral judgment in 6th and 8th graders between Berlin, Bologna, and Warsaw (Silbereisen, Lamsfuss, Boehnke, & Eisenberg, 1991); and a cross-cultural project in the United States and Brazil on levels of adolescents' moral reasoning (Carlo, Koller, Eisenberg, DaSilva, & Frohlich, 1996). Internationalization of research on altruism has been gathering speed in recent decades, and there are both opportunities and an urgent need for studies that would help identify the panhuman and culture-specific prosocial characteristics. Pioneers of the retrospective investigation of altruism amidst the genocidal and other murderous disasters, above all the Holocaust, gathered information regardless of site in documenting the characteristics of heroic altruists. Alas, in the twentieth century, humankind has witnessed a virtual avalanche of national, ideological, racial, religious, and ethnic strife and persecution, and this violence has not ended with the new millennium. Manifestations of altruism have been a ray of hope against the dark background of inhumanity at its extreme. As yet, little is known about the altruistic acts in and around the Gulags of the Soviet Union or during the excesses of the Cultural Revolution in Mao's China, and helpful and humane deeds amidst the eruptions of homicidal fury from Cambodia to Rwanda remain sparsely documented, especially in regard to shelterers and rescuers of potential victims. Moreover, there are the natural disasters to consider, from the tsunamis of South Asia and Japan to Hurricane Katrina and the momentous earthquake in Haiti. All of these catastrophes brought forth tremendous suffering, but also heroism, altruism, and helpfulness that can still be investigated, albeit retrospectively.

1.2.2.7 Interdisciplinary Collaboration

Altruism as a phenomenon is located at the intersection of several disciplines. It is a topic of inquiry by philosophers and theologians, and it is an object of investigation by social, behavioral, and biological scientists. Neuroscience, psychology, sociology, ethnography, and anthropology have altruism in their sights. This diversity is reflected in the orientation and authorship of the chapters in this volume. Its advantages lie in the variety of perspectives and methods brought to bear upon the study of altruism and in their subsequent cross-fertilization; its risks are the potential for fragmentation, isolation, and encapsulation. These dangers can be averted, or at least mitigated, by interdisciplinary communication, and eventually collaboration, which are the objectives that the editor and the contributors to this volume have set out to promote. Alternation of methods and perspectives should be encouraged. For example, a quasi-experimental study of bystander intervention may be followed up by a series of biographical case studies of individuals at both ends of the distribution. A characteristic that may have emerged at the case level in a study of particularly generous philanthropic donors might be more systematically investigated by being incorporated into items of standardized self-report scales and subsequently validated. Hofstede's dimensions, which were identified in the course of a world-wide study of modern nation states, may be explored for their potential relevance in traditional small-scale cultures. Two cultures, for example, may be preselected on the basis of their respective prominently masculine and feminine characteristics and then further scrutinized by qualitative methods in order to ascertain how these features cohere in the various domains of life, particularly in relation to altruism.

1.3 Conclusion

Are human beings "compassionate beasts"? (Hunt, 1990, p. 12) Is altruism a human "quality that dwells at the very heart of reality"? (Penner, 1995, p. 113) There is no way of confirming or refuting these far-reaching assertions, although empirical research has established that the motivational sources of altruism run deep. Altruism appears to be a part of the human potential, perhaps as strong as the much more readily acknowledged and observed human penchant for destructiveness. The collective contribution of social scientists has been in specifying some of the conditions under which altruism impels action and animates conduct. Further interdisciplinary research effort, pursued from a variety of perspectives with multiple methods, holds the promise of hastening the ongoing transition of altruism from a phenomenon to be observed through a construct to be specified to a variable to be investigated.

References

Bar-Tal, D. (1982). Sequential development of helping behavior: A cognitive-learning approach. *Developmental Review, 2*, 101–102.

Batson, C. D. (1991). *The altruism question: Toward a social-psychological answer*. Hillsdale, NJ: Erlbaum.

Batson, C. D. (2010). Empathy-induced altruistic motivation. In M. Mikulincer & P. R. Shaver (Eds.), *Prosocial motives, emotions, and behavior* (pp. 15–35). Washington, DC: American Psychological Association.

Batson, C. D. (2011). *Altruism in humans*. New York: Oxford University Press.

Becker, S. W., & Eagly, A. H. (2004). The heroism of men and women. *American Psychologist, 59*, 163–178.

Bierhoff, H. W. (2001). *Prosocial behavior*. Hove, East Sussex: Psychology Press.

Bierhoff, H. W., Klein, R., & Kramp, P. (1991). Evidence for the altruistic personality from data on accident research. *Journal of Personality, 59*, 263–280.

Boehnke, K. (1988). *Prosoziale Motivation, Selbstkonzept und politische Orientierung: Entwicklungsbedingungen im Jugendalter [Prosocial motivation, self concept, and political orientation: Developmental conditions of change in adolescence]*. Frankfurt: Peter Lang.

Boehnke, K., Silbereiseen, R. K., Eisenberg, N., Reykowski, J., & Palmonari, A. (1989). Developmental pattern of prosocial motivation: A cross-national study. *Journal of Cross-Cultural Psychology, 20*, 219–243.

Bond, M. H., et al. (2004). Culture-level dimensions of social axioms and their correlates across 41 nations. *Journal of Cross-Cultural Psychology, 35*, 548–570.
Buss, D., et al. (1990). International preferences in selecting mates: A study of 37 cultures. *Journal of Cross-Cultural Psychology, 21*, 5–47.
Carlo, G., Koller, S. H., Eisenberg, N., Da Silva, M. S., & Frohlich, C. B. (1996). A cross-national study of the relation among prosocial moral reasoning, gender role orientation, and prosocial moral reasoning, gender role orientation, and prosocial behaviors. *Developmental Psychology, 32*, 231–240.
Cialdini, R. B., Brown, S. L., Lewis, B. P., Luce, C., & Neuburg, S. L. (1997). Reinterpretation of the empathy-altruism relationship: When one into one equals oneness. *Journal of Personality and Social Psychology, 52*, 749–758.
Claar, A., Boehnke, K., & Silbereisen, R. K. (1984). Zur Entwicklung von Motiven prosozialen Handelns bei 12- bis 18-jaehrigen deutschen und polnischen Schuehlern [Development of motives of prosocial action among 12 to 18 year old German and Polish students]. In A. Strikrud (Ed.), *Jugend und Werte –Aspekte einer politischen Psychologie des Jugendalters* (pp. 166–183). Weinheim: Beltz.
Draguns, J. G. (2007). Empathy across national, cultural, and social barriers. *Baltic Journal of Psychology, 8*, 5–20.
Eisenberg, N., Boehnke, K., Schuhler, K., & Silbereisen, R. K. (1985). Development of prosocial behavior and cognition in German children. *Journal of Cross-Cultural Psychology, 16*, 69–82.
Eisenberg, N., Carlo, G., Murphy, B., & Van Court, P. (1995). Prosocial development in late adolescence: A longitudinal study. *Child Development, 66*, 1179–1197.
Eisenberg, N., Cumberland, A., Guthrie, I. K., Murphy, B. C., & Shepard, S. A. (2005). Age changes in prosocial responding and moral reasoning in adolescence and early adulthood. *Journal of Research on Adolescence, 15*, 235–260.
Eisenberg, N., & Miller, P. (1987). Empathy and prosocial behavior. *Psychological Bulletin, 101*, 91–119.
Eisenberg, N., & Strayer, J. (Eds.). (1987). *Empathy and its development*. New York: Wiley.
Feldman, R. E. (1968). Response to compatriot and foreigner who seek assistance. *Journal of Personality and Social Psychology, 10*, 202–214.
Gilbert, M. (2003). *The righteous: The unsung heroes of the Holocaust*. New York: Holt.
Hallie, P. P. (1979). *Lest innocent blood be shed: The story of the village of Le Chambon and how goodness happened there*. New York: Harper and Row.
Hein, G., & Singer, T. (2010). Neuroscience meets social psychology: An integrative approach to human empathy and prosocial behavior. In M. Mikulincer & P. R. Shaver (Eds.), *Prosocial motives, emotions, and behavior* (pp. 109–126). Washington, DC: American Psychological Association.
Hoffman, M. I. (1981). Is altruism part of human nature? *Journal of Personality and Social Psychology, 40*, 121–137.
Hofstede, G. (2001). *Culture's consequences: Comparing values, behaviors, institutions, and organizations across cultures* (2nd ed.). Thousand Oaks, CA: Sage.
Hunt, M. (1990). *The compassionate beast: What science is discovering about the humane side of humankind*. New York: William Morrow and Co.
Kakavoulis, A. (1998). Early childhood altruism: How parents see prosocial behavior in their young children. *Early Child Development and Care, 140*, 115–126.
Knafo, A., & Israel, S. (2010). Genetic and environmental influences on prosocial behavior. In M. Mikulincer & P. Shaver (Eds.), *Prosocial motives, emotions, and behavior* (pp. 149–169). Washington, DC: American Psychological Association.
Kraus, M. W., Piff, P. K., Mendoza-Denton, R., Rheinschmidt, M. L., & Keltner, D. (2012). Social class, solipsism, and contextualism: How the rich are different from the poor. *Psychological Review, 119*, 546–572.
Krebs, D. L. (1975). Empathy and altruism. *Journal of Personality and Social Psychology, 32*, 1134–1146.

London, P. (1970). The rescuers: Motivational hypotheses about Christians who saved the Jews from the Nazis. In J. Macauley & L. Berkowitz (Eds.), *Altruism and helping behavior* (pp. 241–250). New York: Academic Press.
Maner, J. K., Luce, C. L., Neuberg, S. L., Cialdini, R. B., Brown, S., & Sagarin, B. J. (2002). The effects of perspective taking on motivation for helping: Still no evidence for altruism. *Personality and Social Psychology Bulletin, 28*, 1601–1610.
McCauley, C., & Bock, J. G. (2004). Why does violence trump peace building?: Negativity bias in intergroup relations. In Y. T. Lee, C. McCauley, F. Mogghaddam, & S. Worchel (Eds.), *The psychology of ethnic and cultural conflict* (pp. 273–288). Westport, CT: Praeger.
McCrae, R. R., & Allik, J. (Eds.). (2002). *The five factor model of personality across cultures*. New York: Kluwer Academic/Plenum.
Mehrabian, A., & Epstein, N. (1972). A measure of emotional empathy. *Journal of Personality, 40*, 525–543.
Mikulincer, M., & Shaver, P. R. (2005). Attachment security, compassion, and altruism. *Current Directions in Psychological Science, 14*, 33–38.
Monroe, K. R. (1996). *The heart of altruism: Perception of a common reality*. Princeton, NJ: Princeton University Press.
Monroe, K. R. (2002). Explicating altruism. In S. G. Post, G. Underwood, J. P. Schloss, & W. P. Hurlbut (Eds.), *Altruism & altruistic love: Science, philosophy, and religion in dialogue* (pp. 106–122). New York: Oxford University Press.
Oliner, S. P. (2002). Extraordinary acts of ordinary people: Faces of heroism and altruism. In S. G. Post, G. Underwood, J. P. Shloss, & W. B. Hurlbut (Eds.), *Altruism & altruistic love: Science and religion in dialogue* (pp. 123–139). New York: Oxford University Press.
Oliner, S. P., & Oliner, P. M. (1988). *The altruistic personality: Rescuers of Jews in Nazi Europe*. New York: Free Press.
Penner, P. S. (1995). *Altruistic behavior: An inquiry into motivation*. Amsterdam: Rodopi.
Piliavin, J. A., & Charng, H. V. (1990). Altruism: A review of recent theory and research. *American Sociological Review, 16*, 27–65.
Post, S. G. (2003). Profiles of unlimited love: Lives ennobled by purpose. In S. G. Post, B. Johnson, M. E. McCollough, & J. P. Schloss (Eds.), *Research on altruism & love* (pp. 183–215). Philadelphia: Templeton Foundation Press.
Rohner, R. P. (2004). The parental acceptance-rejection syndrome. *American Psychologist, 59*, 830–849.
Royzman, E. & Kumar, B. (2001). On the relative preponderance of empathetic sorrow and its relation to commonsense morality. *New Ideas in Psychology, 19*, 131–144.
Rushton, J. R. (1984). The altruistic personality: Evidence from laboratory, naturalistic, and self-report. In E. Staub, D. Bar-Tal, J. Karylowski, & J. Reykowski (Eds.), *The development and maintenance of prosocial behavior: International perspectives on positive morality* (pp. 271–290). New York: Plenum Press.
Rushton, J. R., Fulker, D. W., Neal, M. C., Nias, D. K., & Eysenck, H. J. (1986). Altruism and aggression: The heritability of individual differences. *Journal of Personality and Social Psychology, 50*, 1192–1198.
Schmitt, D. P., et al. (2004). Patterns and universals of romantic attachment across 62 regions. *Journal of Cross-Cultural Psychology, 26*, 92–116.
Schwartz, S. H. (2010). Basic values: How they motivate and inhibit prosocial behavior. In M. Mikulincer & P. R. Shaver (Eds.), *Prosocial motives, emotions, and behavior* (pp. 221–241). Washington, DC: American Psychological Association.
Silbereisen, R., Lamsfuss, S., Boehnke, K., & Eisenberg, N. (1991). Developmental patterns and correlates of prosocial motives in adolescence. In L. Montada & H. W. Bierhoff (Eds.), *Altruism in social systems* (pp. 82–104). Lewiston, NY: Hogrefe.
Sommerfeld, E. (2010). The subjective experience of generosity. In M. Mikulincer & P. R. Shaver (Eds.), *Prosocial motives, emotions, and behavior* (pp. 303–324). Washington, DC: American Psychological Association.

Sorokin, P. A. (1950). *Altruistic love: A study of American "good neighbors" and Christian saints.* Boston: Beacon Press.
Staub, E. (2003). *The psychology of good and evil: Why children, adults, and groups help and harm others.* New York: Cambridge University Press.
Stotland, E., Mathews, K. E., Sherman, S. E., Hanson, R. O., & Richardson, B. Z. (1978). *Empathy, fantasy, and helping.* Beverly Hills, CA: Sage.
Svetlova, M., Nichols, S. R., & Brownell, C. A. (2010). Toddlers' prosocial behavior: From instrumental to empathetic to altruistic helping. *Child Development, 81*, 1814–1827.
Tec, N. (1986). *When light pierced the darkness.* New York: Oxford University Press.
Tolmacz, R. (2010). Forms of concern: A psychoanalytic perspective. In M. Mikulincer & P. R. Shaver (Eds.), *Prosocial motives, emotions, and behavior* (pp. 93–108). Washington, DC: American Psychological Association.
Vaillant, G. E. (1977). *Adaptation to life.* Boston: Little, Brown.
Vollhart, J. R., & Staub, E. (2011). Inclusive altruism born of suffering: The relationship between adversity and prosocial attitudes and behavior toward disadvantaged outgroups. *The American Journal of Orthopsychiatry, 81*, 307–315.
Zahn-Wexler, C., Radke-Yarrow, M., Wagner, E., & Chapman, M. (1992). Developmental concern for others. *Developmental Psychology, 28*, 126–136.

Chapter 2
Conceptual Aspects of Altruism in Cross-Cultural Contexts

Sandi W. Smith, Maria Knight Lapinski, Mary J. Bresnahan, and Stacy L. Smith

2.1 Introduction

In today's global society, it is important to study and analyze behavioral manifestations of altruism cross-culturally. The ideas discussed here originated with an effort to develop conceptual definitions of altruism and related issues and an operational coding scheme that could be used to code instances of altruistic behavior as it appeared on American television (Smith et al., 2009). As such, the definitions and coding scheme reflected a societal conception rooted in an individualistic culture. The purpose of this chapter is to describe the conceptual definitions and coding scheme and to determine in what ways this can be further extended or modified to reflect communal cultures (Smith, Bresnahan, & Smith, 2011). Therefore, we are reconsidering the concept of altruism in light of possible cultural variability in order to expand research coding the altruistic content of television programs (Smith et al., 2006) to include behavior in different contexts and cultures. The coding of such content is challenging because it involves examining the context and messages exchanged among actors to determine the extent to which they embody the dimensions outlined below. This type of coding scheme has relevance for understanding the communication of altruism in verbal exchanges of all types, including interviews, policy debates, negotiations, physician-patient communication, and relational communication, among other contexts. As Arkhipova and Kozmin (2013) demonstrate in Chap. 5 of this volume, coding content of narratives such as fairy tales and animal

S.W. Smith (✉) • M.K. Lapinski • M.J. Bresnahan
Department of Communication, Michigan State University,
404 Wilson Road, East Lansing, MI 48824, USA
e-mail: smiths@msu.edu; lapinsk3@msu.edu

S.L. Smith
Annenberg School for Communication, University of Southern California,
3502 Watt Way, Los Angeles, CA 90089-0281, USA
e-mail: stacy.smith@usc.edu

D.A. Vakoch (ed.), *Altruism in Cross-Cultural Perspective*,
International and Cultural Psychology, DOI 10.1007/978-1-4614-6952-0_2,

tales can provide insight into the ways in which different ethnic cultures value particular characteristics, including altruism. By expanding the conceptualization of altruism, which is the basis for this coding, we hope to further work on altruism to be more inclusive of multiple cultural perspectives.

Smith et al. (2006) created a conceptual definition of altruistic behavior as "behavior that is intended to benefit others beyond simple sociability or duties associated with role (i.e., family or work). All altruistic behaviors, by definition, must be legal" (p. 711). More specifically, we defined altruistic behavior as acts of cognitive or physical helping and sharing (including giving and donating) that occur outside the bounds of role relationships. Thus, all altruistic actions associated with the *normal* duties of an occupation or social role were not included in that definition. However, there are times when people operating within the bounds of a particular occupation or role go above and beyond normal expectations. To illustrate, a first responder may run into an inferno to save a pregnant mother despite the fact that all of his training would suggest not doing so. A doctor may read to an unconscious child in intensive care hours after performing surgery for a congenital heart defect. Such acts are not only attempts at benefiting another but also defy or exceed norms associated with a particular role. Such acts would also be considered acts of altruism in communal cultures, as they are clearly identified as voluntary (see Chap. 6 of this volume by Grönlund (2013), for additional discussion) and intentional actions that go beyond the boundaries of role-related norms.

However, communal cultures are often based on more fully articulated systems of social obligations to others compared to more individualistically oriented cultures (Bresnahan, 1991; Clark & Mills, 1993; Janoff-Bulman & Leggatt, 2002; Miller & Bersoff, 1998). Communal or collectivistic cultural values emphasize connectedness within in-groups and place higher priority on the goals, needs, and concerns of the group over those of individuals (e.g., Hofstede, 2001; Triandis, 1995). Individualistic cultural values emphasize autonomy and personal choice and place higher priority on the goals, needs, and concerns of the self over those of the group. There is some evidence that these preferences can influence self-conceptions and as such can be assessed at an individual level (Markus & Kitayama, 1991). It is clear that altruistic behavior has some basis in both culture and genetics (see Chap. 3 of this volume by Coe and Palmer (2013)).

The distinction between communal and individualistic cultures was chosen as a point of focus here, despite the multiple ways one might distinguish among cultures, for several key reasons. First, this distinction focuses on the relationship of the self to others within a society and is inherently relational in nature; this is critical when considering altruistic behaviors. Indeed, theories that address the basis of altruistic behaviors suggest culture may drive altruistic behaviors along with genetically-based kinship and reciprocity explanations (e.g., see Moody, 2008). Second, the distinction between a communal and individualistic cultural orientation has been linked to the ways in which people communicate and their social-psychological experience (see Kim, Aune, Hunter, Kim, & Kim, 2001; or Lapinski, Rimal, DeVries, & Lee, 2007). The analysis of altruism here focuses on the ways in which altruism is communicated through overt behavior.

Although on the surface many exchanges between people appear to be altruistic, they may be motivated and explained by processes such as the maintenance of face relations and obligatory exchange rather than altruism (Ting-Toomey & Chung, 2005). Indeed the anticipation of experiencing guilt (Lindsey, 2005), shame, and other negative emotions as a result of unfulfilled role obligation (Bedford & Hwang, 2003) or normative violations (Bierbrauerin, 1992) may drive what appear on the surface to be altruistic behaviors. Thus, the boundary between obligatory role-expected behavior and voluntary action is often blurred in more communal cultures, particularly for the outside observer. An important goal here is to refine the proposed coding scheme for interpreting acts of altruism to be sensitive to such important cultural differences. In order to do so, we first conceptualize altruism along five key dimensions.

2.2 Conceptualizing Altruism

In 1851, Auguste Comte penned the term "altruism." Derived from the Latin word "alter" (i.e., meaning other) and the Italian adjective *altrui*, Comte (1875) believed that altruism signified benevolence or living for others. Decades have passed and much debate has ensued since Comte originally defined the term and its selfish counterpart, egoism. Now there is much disagreement on the limiting conditions surrounding altruistic acts (see Post, Underwood, Scholss, & Hurlbut, 2002).

We believe that the variability in definitions is something to embrace rather than eschew; definitions, after all, are best evaluated in terms of their utility as opposed to some objective assessment of quality. Instead of wrestling over what constitutes an "altruistic act" outside a voluntary action that is intended to benefit others beyond simple sociability or duties associated with role, we have operationalized aspects of different definitions offered in the literature so as to embrace different *researchers'* conceptualizations of the construct.

After reviewing the literature, five key aspects of altruism came to the fore that some scholars include and some exclude in their conceptual definitions. It is important to note that much of this literature is grounded in theorizing that is yet to be tested empirically. Further, this work is based largely on work conducted from Western perspectives and has focused largely on human-human altruism as opposed to altruistic actions directed toward other living things (termed "biosphere altruism" by Gupta, 2013 in Chap. 8 of this volume), which may inhibit the generalizability of the ideas postulated here even though we have intentionally worked to consider the literature on cultural variability in our thinking. In addition to what we review below, two other attributes—voluntary and intentional—are generally agreed upon in the literature and were captured in our basic definition of altruistic behavior (see Monroe, 2002; Oliner, 2002). Thus, these attributes will not be reviewed below, although in communal cultures defining voluntary behavior is somewhat more difficult than in an individualistic culture. To a lesser extent intentionality could be problematic as well. These attributes should be present in all altruistic behavior, no matter the culture, even from the most liberal definition of the term. Five components of altruistic behavior are addressed here, and three of these are predicted to differ according to the culture in which they are enacted: determining whether the act was motivated by a primary *concern for the other*, whether the actor would be likely to engage in self-blame or be socially censured if he or she did not engage in the action (termed "*ease of escape*"), and *actual benefits to the receiver*. The other two components of altruistic behavior we postulate to operate pan-culturally are *empathy* and *cost to the initiator*.

2.2.1 Concern

One of the common definitional elements of altruism centers on individuals' locus of concern when performing altruistic acts. Some theorists have argued that the primary concern of the altruist is for the other (Batson, 2002; Eisenberg & Miller, 1987; Kagan, 2002; Latane & Darley, 1970; Oliner, 2002; Rushton, 1976) and not the self (see Monroe, 2002, p. 107). For example, Post (2002) argues that, "By the strictest definition, the altruist is someone who does something for the other and for the other's sake, rather than as a means to self-promotion or internal well being" (p. 53). Whether we label the motivational state a "goal" or "concern," theorists are arguing that the primary intent behind helping behavior is to facilitate an "other" over self in some way. Theorists also have been quite clear that there may be secondary concerns (i.e., motivational pluralism) associated with altruistic acts (Post, 2002, p. 53; Sober, 2002, p. 19). For example, a young boy may rescue a scared and injured dog that had accidentally fallen into a storm drain. After rescuing the animal and trying to find its owner, the boy may wonder whether he will get a reward for the dog's return. Such self-motivated concerns, provided that they are not the primary reason for performing such an act, do not disqualify the act from being altruistic in nature for some altruism theorists. Types of self-concerns may include, but are not limited to, self-promotion, internal well-being, alleviation of a negative state, positive self-face, and avoidance of punishment (see Batson, 2002; Post, 2002).

To capture these issues, the coding scheme must include a variable assessing whether benefit to the self or other is the primary force behind a character's decision to act altruistically. Therefore, coders are trained to decipher concern based not only on verbal utterances made by the initiator of the act but also their nonverbal responses and the context of the unfolding situation. The difficulty in coding this construct lies in determining the point at which concern for other is higher than concern for self. This is particularly difficult in cultures and situations in which there are strict, but implicit, social obligations to others. If an actor engages in an act primarily to avoid censure, shame, and punishment, the concern was higher for self even though it might seem primarily to benefit the other on the surface. Another way to understand this issue is to consider the relational context in which an act is taking place. In communal cultures, role obligations, for example, are thought to be bound to in-group members. As such, acts which show concern for out-group members may not be motivated by social obligation but by altruism.

Therefore, knowledge of the demands of the culture is critical when coding action as concern for other or self. Each action can be coded, ultimately, as either primary concern for self or primary concern for other. This apparent problem of deciding whether an action reflects self or other concern can be resolved by having cultural insiders from the relevant culture who are intimately familiar with the obligation system of that culture do the coding based on indigenous values present in the cultural system. In collaboration with representatives of the target culture being studied, the coding system can be amended where needed to be sensitive to the demands of another value system while the basic framework of the coding system is maintained.

2.2.2 Cost

Often, the word altruism conjures up extreme images of individuals risking life and limb for the sake of saving another from the hands of death. Central to this idea is the belief that altruistic acts involve a sacrifice or cost on the part of the initiator. Several theorists hold this view (Monroe, 2002; Sober, 2002; Wyschogrod, 2002). Oliner (2002), one of the most notable sociologists in this area, arranges costly altruism on a continuum from heroic acts to more conventional daily experiences.

Many of Oliner's ideas about altruism are derived from hundreds of interviews with rescuers of Jews during the time of the Holocaust (see Oliner & Oliner, 1988). The researchers found that many individuals risked not only their own lives but also the lives of family and friends in an effort to save those destined to death. Some of the instances involved single, extraordinary acts of heroism that saved lives, whereas other efforts involved extended acts of giving and hospitality (i.e., hiding Jews in their home) in the continued face of fear. Similar results were found in Monroe's (1996) study, which involved a substantially smaller sample of interviews with rescuers of Jews from World War II.

In an effort to measure this aspect of altruism, we created a variable designed to determine whether the initiator experiences a "cost" for helping another. Costs are defined broadly and may be physical (i.e., injury/death), emotional (i.e., embarrassment, grief), and/or material (i.e., loss of home, money, car) in nature. Given that altruism encompasses all types of "costs" ranging from the tragic to the trivial, the variable captures only the presence or absence of a cost and does not ask coders to determine the degree or intensity of the potential loss on some sort of scale. The idea of altruistic acts as sacrifice or cost on the part of the initiator is theorized to be a pan-cultural phenomenon, and while the degree of cost will likely vary by culture and context, the concept represented by cost should characterize altruistic behaviors in all cultures (see Chap. 11 of this volume by Chick (2013), for an example of costs and benefits associated with the *cargo* system in Latin America).

2.2.3 *Benefit to the Recipient*

A logical extension to an act that is motivated by concern for the other and that is costly to the actor is the fact that the recipient should actually benefit from the act. Some scholars argue that altruism cannot occur without actual benefit accruing to the recipient as a result of the act.

Recipient benefit refers to something that actually promotes or enhances the life of the recipient. Benefits may be emotional (i.e., confidence, self-esteem), physical (i.e., ability to walk), material (i.e., car, house), or spiritual (i.e., faith) in nature. Each act can be coded as recipient benefit present or absent.

Similar to the case of coding the locus of concern, a case can be made that benefits should accrue to only the recipient rather than to the initiator when assessing recipient benefit. The notion of who benefits from an action becomes harder to distinguish when one considers the possibility of self-conceptualizations in which the self is inextricably tied to others. Sampson (1988) and others (Markus & Kitayama, 1991) have addressed the concept of the interdependent self-construal, thought to predominate in communal cultures, in which the self is defined only in relationship to others. In this case, benefits to a recipient should also accrue to the actor. That is, if an individual engages in actions to benefit others and sees those others as fundamentally tied to his or her own identity, they will benefit individually from improvement in the well-being of the other. This should be the case only for people with whom one has a shared sense of identity such as in-group members. Again, the relational context may help to explain whether or not an act truly benefits only the recipient and not the actor, if the person is someone with whom one does not likely have shared self-identification (e.g., a complete stranger) and the benefits should occur for the recipient independent of the actor alone.

2.2.4 *Empathy*

It has been argued that one of the reasons individuals' help distressed others may be because of empathy (Batson, 2002). In fact, several studies have found that empathy evoked by witnessing others in distress facilitates helping behavior (Batson, Duncan, Ackerman, Buckley, & Birch, 1981; Coke, Batson, & McDavis, 1978; Fultz, Batson, Fortenbach, McCarthy, & Varney, 1986). Yet, meta-analyzes reveal that the strength of the relationship may vary depending on the operationalization of altruistic action, the method of measuring empathy (self-report, picture indices, physiological markers), and age of the participant in the study (Eisenberg & Miller, 1987).

Empathy is a controversial construct in the social science literature that is defined in multiple ways (Eisenberg & Miller, 1987; Feshbach & Feshbach, 1997; Hurlbut, 2002; Zillmann, 1999). Some researchers define empathy in terms of affect matching (i.e., facial mimicry), emotional responding (i.e., sharing the same or similar emotional state), cognitive reactions (i.e., ability to take the perspective of the other, concern for other's plight), and/or some combination of these categories.

Most scholars agree that cognitive and affective factors are both at work in empathic reactivity (see Eisenberg & Strayer, 1987; Feshbach & Feshbach, 1997). The most extreme reaction to another's need is emotional contagion, whereby the individual not only perceives the need in the other but is so overwhelmed by the emotion that it becomes self-, not other, focused (Preston & DeWaal, 2002).

Given this literature, three measures arise which assess different approaches to empathy. The coding scheme uses dichotomous variables, to examine whether the initiator (1) has the capacity to take the perspective of the character in need, (2) shows empathic concern for the other, and (3) becomes self-focused in his/her emotional responsiveness (i.e., contagion effect). These measures are combined so that empathy occurs when "1" and "2" are present but "3" is absent. For example, very young children might not have developed the capacity to take the perspective of the other even though they might exhibit behavior that appears to show empathic concern. Other behaviors might benefit the recipient but not be offered out of a spirit of helping. Finally, an initiator might be overwhelmed with sadness for the recipient to the extent that he or she is unable to act. In each of these cases, empathic behavior would be absent.

It is important to note, however, that it is altogether possible that measuring internal cognitive states such as perspective taking may be impossible to ascertain from behavior. Even if an actor seems to have the "capacity" or shows signs of perspective taking, it may be impossible to know if this is in fact what she/he is doing. Once again, we believe that this aspect of altruistic behavior should operate pan-culturally.

2.2.5 *Ease of Escape*

A fifth central feature of the empathy-altruism hypothesis is the notion of ease of escape (see Batson, 2002). Very simply, Batson (1991, 2002) has argued and experimentally tested other motives that might drive helping behaviors such as aversive arousal, reward seeking, or punishment avoidance. All are considered egoistic in nature; the basic premise is that when empathy is low or nonexistent, any one of these other self-focused motives may drive positive social actions such as sharing, giving, or donating. He has tested these egoistic alternatives with the variable "ease of escape," or the relative effort it takes one to withdraw from potential helping situations.

Ease of escape is defined operationally in two ways. The first is self-blame or internal negatively valenced emotions such as guilt or shame (Bedford & Hwang, 2003; Bierbrauerin 1992; Lindsey, 2005; O'Keefe, 2002). Typically, these are punishment-based feelings that might emerge in the face of not helping another in need. Ease of escape is high if one can remove the self from the potential helping situation without feeling bad, guilty, or remorseful in some way. Thus, the inability to escape in the absence of empathy might suggest that one is helping to reduce aversive arousal or internal punishment within. In communal cultures, the inability of escape ties in closely to the obligation system. Even when obligation is relatively low, there may be negative self-consequences from failure to help a member of one's in-group. It is important to point out that different standards may apply to interactions with members of an out-group, especially in communal cultures, but this distinction might not be found in more individualist cultures. Thus, internally driven ease of escape, such as helping to avoid anticipated guilt or shame, may be another area where cultural variance can be expected.

The second operational definition of ease of escape is social censure or external factors that may evoke condemnation from others. Ease of external escape occurs if one can remove the self from the potential helping situation without enduring the condemnation of others for failing to help or somehow avoid others knowing that you failed to help. The inability to escape the helping scene—in the absence of empathy—might suggest that one is helping to avoid social punishment or to receive rewards from bystanders.

We believe external ease of escape may vary depending on whether a culture is communal or individualistic. For example, avoidance of social sanctions may be less possible in communal cultures where social norms tend to have a stronger influence (Bagozzi, Wong, Abe, & Bergami, 2000; Park & Levine, 1999) relative to individualistic cultures. Because the power of actual or anticipated social sanctions has a greater influence on behaviors in communal cultures, the extent to which people believe it is possible to escape these sanctions is decreased, thus motivating helping behaviors which in individualistic cultures are considered altruistic. Actions done in private, without the possibility that others might have direct or indirect knowledge of the behavior, are likely to enhance the potential for external ease of escape by decreasing the probability of social sanctions (Lapinski & Rimal, 2005).

Studies have typically found that egoistic motivations for helping operate in the absence of empathy (for excellent review, see Batson, 2002). The two measures in the present research that capture ease of escape are the presence or absence of internal blame (i.e., self-censure) and external blame (i.e., other censure) for each helping incident. Taken in combination with the empathy measure outlined above, the ease of escape variables helps to ascertain egotistic reasons for helping when empathic reactivity is not present.

In sum, five different variables capture differences in altruistic actions in our conceptualization. They can be used to create different composites that reflect different definitions of altruism. Cost and empathy seem to bridge communal and individualistic cultures, whereas concern for other versus self, benefit to recipient, and ease of escape are critical to understanding the differences in altruism between cultures.

2.3 Composites of Altruism

Due to the aforementioned ambiguity surrounding the conceptual definition of altruism, we created four specific composites of altruism. See Table 2.1 for an overview of the variables and composites that result from grouping them as described below. The first composite simply involved instances of helping and/or sharing.

Table 2.1 Composite definitions of altruism

	Liberal composite	Initiator focus	Recipient focus	Altruistically loving behavior
Helping/sharing	Yes	Yes	Yes	Yes
Cost		Yes		Yes
Concern for other		Yes		Yes
Benefit for other			Yes	Yes
Empathy			Yes	Yes
Internal and external ease of escape				Yes

No additional stipulations were added to these types of acts, which should render this the most *liberal* composite of altruism.

The second and third composites were informed by the work of Krebs (1970). Arguing for a framework of altruism, Krebs (1970) asserts, "To begin with, the prototypical altruistic situation involves someone who gives (a benefactor), and someone who receives (the recipient). In some cases, characteristics of the benefactor affect altruism, and in other cases it is characteristics of the recipient. The first dimension of classification, then, separates variables which relate to the characteristics of benefactors that cause or correlate with altruism from the altruism-eliciting characteristics of recipients" (p. 262). Using Krebs' (1970) logic, the second composite tapped key variables related to the *initiator* of altruistic acts such as the locus of concern and cost. The second composite includes instances of helping and/or sharing that were motivated out of a primary concern for the other over self and involved personal cost to the initiator. In cross-cultural applications of this coding scheme, ascertaining the locus of concern will be of primary importance.

The third composite tapped key variables related to the *recipient* and featured acts of helping/sharing that benefited the recipient and were the by-product of empathy. These acts were motivated by initiator projection into the emotional state and need of the recipient so that he or she could act in such a way that actually benefited the recipient. The extent to which recipient benefit is independent of actor benefit should be more carefully examined in communal relative to independent cultures.

The fourth composite is the most conservative. Only acts of helping/sharing that feature all five dimensions were included. These are instances in which the initiator is primarily concerned with the other, there is a cost to the actor, the recipient actually benefits, the act is the by-product of empathy, and ease of escape from self-censure or social censure is available. It is our belief that this stringent composite captures the most conservative other-oriented instances of altruistic behavior. Such acts have been described in the literature to be on par with the actions of receivers of the Carnegie Hero Commission Award, hospice volunteers, rescuers of Jews in Nazi Europe during World War II, and similar heroic actions (Monroe, 2002, p. 108; Oliner, 2002, pp. 123–133). The escape from internal censure in this strict definition will be a variable that might differ cross-culturally, as will the primary concern for other over self and the locus of benefit variables.

In sum, we defined altruistic acts as instances of helping and sharing. We presented five variables that theorists identify as critical components of altruism, and we created from these variables four composite definitions of altruism. The "purest" form of altruism may involve those altruistic acts that stem from a primary concern for the other, actually benefit the recipient, involve empathy, incur a cost to the initiator, and from which the actor could escape self-censure or social censure relatively easily. These acts might be termed *altruistically loving behavior*. More liberal forms of altruism may include only one, two, or three of these elements, which would be more consistent with conceptualizations of this construct by Batson (2002), Oliner (2002), and Monroe (2002).

The different composites of altruism all are likely to be coded differentially across communal and individualistic cultures. The behavioral component that is the bedrock of all of the composites, helping and sharing, might be motivated more often by role-related expectations in communal cultures. In addition, the initiator component of primary concern for the other, the recipient component of benefit, and the altruistic love component of internal ease of escape are predicted to differ in communal versus individualistic cultures.

While we believe there is pan-cultural commonality in the meaning of acts of altruism, we have identified the elements in our proposed coding system that we believe will be most susceptible to cultural variation. We have proposed that as this coding scheme is extended to other cultures, modifications must be crafted by working in tandem with collaborating scholars from these target cultures who will be able to provide guidance on the cultural values that need to be factored into our coding scheme. In particular, we suggest that a critical need exists to accurately reflect differences in interpersonal obligation and mutual face needs as they relate to whether the act was beyond role expectations, the primary force motivating the behavior was concern for the other, and variability in ease of escape from one's internalized obligations to others in the in-group and the out-group in order to be able to code cross-cultural altruistic behavior.

Acknowledgments This work was supported by a grant from the Fetzer Institute to Sandi W. Smith and Stacy L. Smith, Principal Investigators. This chapter is an adaptation of Smith et al. (2011). Application of the altruistic behavior coding scheme to cross-cultural contexts. *World Cultures eJournal, 18*(1). It is republished in this volume with permission.

References

Arkhipova, A., & Kozmin, A. (2013). Do we really like the kind girls and animals?: Cross-cultural analysis of altruism in folktales. In D. A. Vakoch (Ed.), *Altruism in cross-cultural perspective*. New York: Springer.

Bagozzi, R. P., Wong, N., Abe, S., & Bergami, M. (2000). Cultural and situational contingencies and the theory of reasoned action: Application to fast food restaurant consumption. *Journal of Consumer Psychology, 9*, 97–106.

Batson, C. D. (1991). *The altruism question: Towards a social psychological answer*. Hillsdale, NJ: Lawrence Erlbaum.

Batson, C. D. (2002). Addressing the altruism question experimentally. In S. G. Post, L. G. Underwood, J. P. Schloss, & W. B. Hurlbut (Eds.), *Altruism and altruistic love: Science, philosophy, and religion in dialogue* (pp. 89–105). New York: Oxford University Press.

Batson, C. D., Duncan, B. D., Ackerman, P., Buckley, T., & Birch, K. (1981). Is empathic emotion a source of altruistic motivation. *Journal of Personality and Social Psychology, 40*, 291–302.

Bedford, O., & Hwang, K.-K. (2003). Guilt and shame in Chinese culture: A cross-cultural framework from the perspective of morality and identity. *Journal for the Theory of Social Behavior, 33*(2), 127–144. doi:10.1111/1468-5914.00210.

Bierbrauerin, G. (1992). Reactions to violation of normative standards: A cross-cultural analysis of shame and guilt. *International Journal of Psychology, 27*(2), 181–193.

Bresnahan, M. I. (1991). *Finding our feet: Understanding cross-cultural discourse*. Lanham, MD: University Press of America.

Chick, G. (2013). Altruism in human ritual. In D. A. Vakoch (Ed.), *Altruism in cross-cultural perspective*. New York: Springer.

Clark, M. S., & Mills, J. (1993). The difference between communal and exchange relationships: What it is and is not. *Personality and Social Psychology Bulletin, 19*, 684–691.

Coe, M. K., & Palmer, C. T. (2013). Cross-cultural variation in altruism: Traditional parental manipulation and ancestor–descendant conflict. In D. A. Vakoch (Ed.), *Altruism in cross-cultural perspective*. New York: Springer.

Coke, J. S., Batson, C. D., & McDavis, K. (1978). Empathic mediation of helping: A two-stage model. *Journal of Personality and Social Psychology, 36*, 752–766.

Comte, A. (1875). *System of positive polity* (Vol. 1). London: Longmans, Green, & Co.

Eisenberg, N., & Miller, P. A. (1987). The relation of empathy to prosocial and related behaviors. *Psychological Bulletin, 101*(1), 91–119.

Eisenberg, N., & Strayer, J. (1987). Critical issues in the study of empathy. In N. Eisenberg & J. Strayer (Eds.), *Empathy and its development* (pp. 3–16). New York: Cambridge University Press.

Feshbach, N. D., & Feshbach, S. (1997). Children's empathy and the media: Realizing the potential of television. In S. Kirschner & D. A. Kirschner (Eds.), *Perspectives on psychology and the media* (pp. 3–28). Washington, DC: American Psychological Association.

Fultz, J., Batson, C. D., Fortenbach, V. A., McCarthy, P. M., & Varney, L. L. (1986). Social evaluation and the empathy-altruism hypothesis. *Journal of Personality and Social Psychology, 50*(4), 761–769.

Grönlund, H. (2013). Cultural values and volunteering: A cross-cultural perspective. In D. A. Vakoch (Ed.), *Altruism in cross-cultural perspective*. New York: Springer.

Gupta, A. (2013). Altruism in Indian religions: Embracing the biosphere. In D. A. Vakoch (Ed.), *Altruism in cross-cultural perspective*. New York: Springer.

Hofstede, G. (2001). *Culture's consequences: Comparing values, behaviors, institutions, and organizations across nations* (2nd ed.). Thousand Oaks, CA: Sage.

Hurlbut, W. B. (2002). Empathy, evolution, and altruism. In S. G. Post, L. G. Underwood, J. P. Schloss, & W. B. Hurlbut (Eds.), *Altruism and altruistic love: Science, philosophy, and religion in dialogue* (pp. 309–327). New York: Oxford University Press.

Janoff-Bulman, R., & Leggatt, H. K. (2002). Culture and social obligation: When "shoulds" are perceived as "wants.". *Journal of Research in Personality, 36*, 260–270.

Kagan, J. (2002). Morality, altruism, and love. In S. G. Post, L. G. Underwood, J. P. Schloss, & W. B. Hurlbut (Eds.), *Altruism and altruistic love: Science, philosophy, and religion in dialogue* (pp. 40–50). New York: Oxford University Press.

Kim, M. S., Aune, K., Hunter, J. E., Kim, H. J., & Kim, J. S. (2001). The effect of culture and self-construals on predispositions toward verbal communication. *Human Communication Research, 27*, 382–408.

Krebs, D. L. (1970). Altruism–an examination of the concept and a review of the literature. *Psychological Bulletin, 73*, 258–302.

Lapinski, M. K., & Rimal, R. N. (2005). An explication of social norms. *Communication Theory, 15*, 127–147.

Lapinski, M. K., Rimal, R., DeVries, R., & Lee, E. L. (2007). The role of group orientation and descriptive norms on water conservation attitudes and behaviors. *Health Communication, 22*, 133–142.
Latane, B., & Darley, J. M. (1970). *The unresponsive bystander: Why doesn't he help?* New York: Appleton-Croft.
Lindsey, L. L. M. (2005). Anticipated guilt as behavioral motivation: An examination of appeals to help unknown others through bone marrow donation. *Human Communication Research, 31*, 453–481.
Markus, H. R., & Kitayama, S. (1991). Culture and the self: Implications for cognition, emotion, and motivation. *Psychological Review, 98*, 224–253.
Miller, J. G., & Bersoff, D. M. (1998). The role of liking in perceptions of the moral responsibility to help: A cultural perspective. *Journal of Experimental Social Psychology, 34*, 443–469.
Monroe, K. R. (1996). *The heart of altruism: Perceptions of a common humanity*. Princeton, NJ: Princeton University Press.
Monroe, K. R. (2002). Explicating altruism. In S. G. Post, L. G. Underwood, J. P. Schloss, & W. B. Hurlbut (Eds.), *Altruism and altruistic love: Science, philosophy, and religion in dialogue* (pp. 106–122). New York: Oxford University Press.
Moody, M. (2008). Serial reciprocity: A preliminary statement. *Sociological Theory, 26*(2), 130–151. doi:10.1111/j.1467-9558.2008.00322.x.
O'Keefe, D. J. (2002). Guilt as a mechanism of persuasion. In J. P. Dillard & M. Pfau (Eds.), *The persuasion handbook: Developments in theory and practice* (pp. 329–344). Thousand Oaks, CA: Sage.
Oliner, S. P. (2002). Extraordinary acts of ordinary people: Faces of heroism and altruism. In S. G. Post, L. G. Underwood, J. P. Schloss, & W. B. Hurlbut (Eds.), *Altruism and altruistic love: Science, philosophy, and religion in dialogue* (pp. 123–139). New York: Oxford University Press.
Oliner, S. P., & Oliner, P. M. (1988). *The altruistic personality: Rescuers of Jews in Nazi Europe*. New York: Free Press.
Park, H.S., & Levine, T. R. (1999). The theory of reasoned action and self-construal: Evidence from three cultures. *Communication Monographs, 66*, 199–218.
Post, S. G. (2002). The tradition of agape. In S. G. Post, L. G. Underwood, J. P. Schloss, & W. B. Hurlbut (Eds.), *Altruism and altruistic love: Science, philosophy, and religion in dialogue* (pp. 51–64). New York: Oxford University Press.
Post, S. G., Underwood, L. G., Scholss, J. P., & Hurlbut, W. B. (Eds.). (2002). *Altruism and altruistic love: Science, philosophy, and religion in dialogue*. Oxford: New York.
Preston, S. D., de Waal, F. B. M. (2002). Empathy: Its ultimate and proximate bases. *Behavioral and Brain Sciences, 25*(1), 1–71.
Rushton, J. P. (1976). Socialization and the altruistic behavior of children. *Psychological Bulletin, 83*(5), 898–913.
Sampson, E. E. (1988). The debate on individualism. *American Psychologist, 43*, 15–22.
Smith, S. W., Bresnahan, M. J., & Smith. S. L. (2011). Application of the altruistic behavior coding scheme to cross cultural contexts. *World Cultures eJournal, 18*(1). Retrieved from http://128.48.120.222/uc/item/4qf127hb.
Smith, S. L., Smith, S. W., Pieper, K. M., Downs, E., Yoo, J., Ferris, A., Bowden, B., & Butler, M. C. (2009). Measuring prosocial behavior, altruism, and compassionate love on US Television. In B. Fehr, S. Sprecher, & L. Underwood (Eds.), *The science of compassionate love: Research, theory, and application* (pp. 53–77). Malden, MA: Blackwell Publishers.
Smith, S. W., Smith, S. L., Pieper, K. M., Yoo, J. H., Ferris, A. L., Downs, E., & Bowden, B. (2006). Altruism on American television: Examining the amount of, and context surrounding acts of helping and sharing. *Journal of Communication, 56*(4), 707–727. doi:10.1111/j.1460-2466.2006.00316.x.
Sober, E. (2002). The ABCs of altruism. In S. G. Post, L. G. Underwood, J. P. Schloss, & W. B. Hurlbut (Eds.), *Altruism and altruistic love: Science, philosophy, and religion in dialogue* (pp. 17–29). New York: Oxford University Press.

Ting-Toomey, S., & Chung, L. C. (2005). *Understanding intercultural communication*. Los Angeles, CA: Roxbury Press.
Triandis, H. C. (1995). *Individualism and collectivism*. Boulder, CO: Westview Press.
Wyschogrod, E. (2002). Pythagorean bodies and the body of altruism. In S. G. Post, L. G. Underwood, J. P. Schloss, & W. B. Hurlbut (Eds.), *Altruism and altruistic love: Science, philosophy, and religion in dialogue* (pp. 29–39). New York: Oxford University Press.
Zillmann, D. (1999). Empathy: Affect from bearing witness to the emotions of others. In J. Bryant & D. Zillmann (Eds.), *Responding to the screen: Reception and reaction processes* (pp. 135–168). Hillsdale, NJ: Lawrence Erlbaum.

Chapter 3
Cross-Cultural Variation in Altruism: Traditional Parental Manipulation and Ancestor-Descendant Conflict

Kathryn Coe and Craig T. Palmer

3.1 Introduction

This chapter starts with the assumption, familiar to evolutionary theorists, that to understand the current cross-cultural patterns in human altruism, you must understand how the forms of altruism unique to humans evolved, and to understand this, you must understand that "We were made for a world that has mostly disappeared, . . . a world in which all activities were enmeshed in webs of kinship . . . a world in which things rarely changed much over the course of a lifetime" (Cronk, 1999, p. 119). While our thesis builds on evolutionary thinking, it deviates from other evolutionary explanations of human altruism because we argue that all of the aspects of our ancestral environment just described are the result of traditions (the behaviors of parents replicated by their offspring). Traditions, by definition, kept human behavior from changing much from one generation to the next over many hundreds and even thousands of years. Traditions are also the *only* mechanism that could have produced the large webs of kinship that constituted the social environment of our ancestors. Most importantly, as we argue in this chapter, traditions were essential underpinnings of the altruistic behavior of the individuals who formed the networks of kin that constituted the social environment of our ancestors. Further, many aspects of altruism, or the lacks thereof, found in much of the world today are the result of the diminishing influence of traditions. Evolutionary explanations that

K. Coe (✉)
Department of Public Health, Indiana University-Purdue University,
714 Senate Ave, Indianapolis, IN 46203, USA
e-mail: coek@iupui.edu

C.T. Palmer
Department of Anthropology, University of Missouri-Columbia,
107 Swallow Hall, Columbia, MO 65211-1440, USA
e-mail: PalmerCT@missouri.edu

D.A. Vakoch (ed.), *Altruism in Cross-Cultural Perspective*,
International and Cultural Psychology, DOI 10.1007/978-1-4614-6952-0_3,

ignore the role of traditions in human altruism in any time or place are ignoring what has been a significant influence on human behavior.

We start our support of these propositions by describing the fundamental dichotomy between traditional and nontraditional societies. We then propose that an explanation of these differences requires an understanding of how traditions encouraging altruism toward co-descendants could have produced the altruism that characterizes the webs of kinship that formed the social environments of our ancestors. We will do this by describing how tantalizingly close several evolutionary theorists have come to this realization over the past 40 years. Then we will describe how patterns of altruism change when societies become less traditional and start to include individuals identified as non-kin.

3.2 Traditional Societies and Distant Kin

One of the most incontrovertible facts of human existence for tens of thousands of years before the Neolithic revolution, and in many parts of the world until even more recently, is the intergenerational "retention and duplication" (Campbell, 1975) of human behavior (p. 1106). This "traditionalness" of human behavior was recognized by Kroeber (1948), who observed that "cultures are ... inclined to be persistent ...[e]ven in times of the most radical change and innovation there are probably several times as many items of culture being transmitted from the past as there are being newly devised" (pp. 256–257). Thus, it is not surprising that nearly anywhere you look in the anthropological literature, you will see references to "traditional" societies. The mere use of this term implies that in the midst of the seeming chaos of cultural diversity in the world, there exists a recognizable dichotomy between traditional societies and nontraditional societies. Although this dichotomy is obviously a continuum, we suggest that when terms are carefully defined, it is a useful place from which to approach the cross-cultural study of altruism.

We define traditional societies as those in which cultural behaviors tend to have been copied from ancestors for *many generations*. These copied behaviors include not only the rituals that are stereotyped and repeated from one generation to the next but also the everyday behaviors related to subsistence and, most importantly, social interaction. As all humans lived in traditional societies until the last few thousand years, even long after the development of agriculture, societies still referred to as traditional resemble in fundamental ways those earlier societies (see discussion in Coe, 2003; Palmer, 2010). Such societies typically consist of *individuals identified as being kin to one another by virtue of being perceived as descended from common ancestors*. Nontraditional societies are those in which traditions have been replaced with cultural behaviors copied from people other than ancestors. The earliest forms of nontraditional societies are often referred to as early nation states. Such early states typically strove to include *multiple* kinship-defined traditional societies (e.g., a number of distinct tribes) and thus were vulnerable to splitting along these kinship divisions (van den Berghe, 1979; Salter, 2002). It is in this period that we see the emergence of fundamental changes in altruistic behavior.

The fundamental difference between traditional societies and nontraditional societies involves kinship. However, the change in kinship occurred primarily at genealogical distances far beyond the close kin where behavior is seen as potentially explainable by kin selection. Thus, the fundamental difference in kinship between traditional and nontraditional societies has gone largely unnoticed by mainstream evolutionary explanations of altruism that tend to restrict their conception of kinship to the close genealogical distances relevant to kin selection (i.e., closer than first cousins). To understand the importance of this difference between traditional and nontraditional societies, it is first necessary to fully grasp the importance of *distant* kin in traditional societies.

Although some traditional societies are small, the tradition of passing descent names or other cultural markers (e.g., body decoration) from ancestors to descendants over many generations enables some traditional societies to become very large, as vast numbers of kin become, over many generations, identified explicitly. As van den Berghe and Barash (1977) explain, unilineal descent "can be seen as a cultural adaptation enabling up to millions of people to organize" (p. 404). Among the Tiv, for example, "the whole population of some 800,000 traces descent by traditional genealogical links from a single founding ancestor" (Evans-Pritchard, 1951, p. 29; Keesing, 1975, pp. 32–33).

While traditions that dictate the use of descent names make it possible to *identify* large numbers of individuals as kin, the mere identification of kin is not sufficient to account for altruism toward individuals identified as kin. Other traditions that encourage altruism toward kin are necessary to produce the altruistic social relationships that form these individual kin into networks commonly called a society (Coe, 2003; Palmer & Steadman, 1997). Traditions encouraging such altruism constitute much, it not all, of the moral codes in traditional societies. Santos Granero (1991) reported that tribal people such as the Peruvian Amuesha regularly claim that "'yi' (morality), which promotes such kinship responsibilities as love and generosity," is crucial to the existence and perpetuation of harmonious and enduring social relationships (p. 226). "Immoral" behaviors, in contrast, are those that are "antisocial," demonstrating selfishness or "greediness or meanness" (Santos Granero, 1991, p. 226) in their "disregard for kinship duties and failure in one's duties towards other fellow Amuesha" (Santos Granero, 1991, p. 45). The claim that altruism directed toward kin is a duty supports a suggestion made by Gibbons (see Chap. 4, this volume) that altruism directed toward kin may be seen as an expectation.

Such traditions encouraging altruism toward kin, and originating from the common ancestors of those kin, are apparently a human universal. Most scholars would agree that the practice of having and enforcing behavioral codes is ancient and that the origin of these codes and the system that enforces them were our ancestors, who, "from time immemorial," were the "primitive custodians of the unwritten, uncodified, unclassified rules of conduct" (Rattray, 1929, p. 3). Primitive law was *ancestral*: "All of it [primitive law]," Culwick and Culwick (1935, p. 8) write, "is neither more nor less than the rules of behavior ordained by the ancestors and practiced by them" (Edel & Edel, 1957, p. 87; Sumner, 1907, p. 232). As Sumner (1907) poetically worded this, these systems "contain in themselves the authority of the ancestral ghosts" (p. 232). Leaders were often, if not universally, claimed to be the representatives of the ancestors. Bandalier (1972) wrote that the primary role of the lineage of clan chief is that he is the representative of the ancestors, "who transmits the words of the ancestors to the living, and those of the living to the ancestors" (p. 99). Shamans were also often said to communicate the wishes of dead ancestors to their living descendants (Steadman & Palmer, 1994).

Moral systems often have no justification other than "we do it this way because the old men say it is wiser" (Sun, 1942, p. 268), or "it was the custom of their ancestors" (Tyler, 1881/1960, p. 252), and it is now our "duty" to our ancestors to behave the way they specified (Edel & Edel, 1957; Johnson, 1984; Westermarck, 1912). It is often claimed that the ancestors who gave the rules still participate in social life, rewarding those who obey and punishing those who violate their rules (Santos Granero, 1991), a claim that may be universal in all traditional societies (Steadman, Palmer, & Tilley, 1996). Among the Ndembu, the "moral man" is one who "honours his kinship obligations" and "respects and remembers his ancestors" (Turner, 1979, p. 374), and Turner points out that these "moral values and . . . ethical code . . . would be recognized as valid by all human groups" (Turner, 1979, p. 374). Middleton (1960) sums this central aspect of human altruism by this simple quote from the Lugbara: "the rules of social behaviour are the 'words of our ancestors'" (p. 27). To act morally is one's duty to the ancestors; morals are not justified by a claim that they are just or fair.

Given the claim that moral codes come from ancestors, it is not surprising that the *scope* of moral codes in traditional society is defined by kinship, not geography (Edel & Edel, 1957, p. 16; King, 1972, p. 37). Specific codes often correspond to specific categories of kin (Coe, 1995; Palmer & Steadman, 1997). Birth and descent alone indicate "those who count in it reckoning and take part in its proceedings" (Edel & Edel, 1957, p. 16). Although descent names can be associated with ancestral lands, *birth* (i.e., descent) is what appears to be important because clans and tribes are merely widely spread categories whose members are identified by descent names or other markers and rarely if ever gathered into one geographic "group" (Edel & Edel, 1957; Palmer & Steadman, 1997).

The key aspect of these traditional moral codes is that they consist of rules encouraging individuals to be altruistic toward distant co-descendants as if they were close kin. As Briffault (1931, p. 57) observed, there are rules of "kindness, love, help, and peace applicable to members of our own clan, tribe, or community, the other of robbery, hatred, enmity, and murder to all the rest of the world" (p. 57). Outsiders [non-kin] in traditional "static" (or unchanging) societies are considered to be less than human (Hoebel, 1949; Santos Granero, 1991).

In kinship-based traditional societies many, perhaps most, rules may be unspoken and are transmitted by copying or modeling, or through verbal behaviors (Van Baal, 1981; King, 1972). Even if unspoken, individuals are quite conscious of a high valuation placed on certain behaviors. Children in all societies are educated about behavioral codes and "the specific consequences that will follow if a rule is not obeyed" (Hoebel, 1949, p. 363). This teaching most often was done in the family. Thus, *the transmission of moral codes consists of parents manipulating the behavior of their offspring to be more altruistic toward co-descendants and to replicate this manipulation when interacting with their own offspring in the future.*

The persistent transmission of unwritten moral codes unchanged from one generation to the next requires considerable effort, often including guided practice and ritualized memorization. Ironically, although writing makes it easier to maintain codes unchanged, it was the written legal codes of early states that often underwent rapid change because they allowed legislative enactments (Diamond, 1951; Wines, 1858, p. 79). In contrast, the unwritten moral codes of traditional kinship societies were passed through both verbal and nonverbal behaviors with little change (Van Baal, 1981; King, 1972). "Numerous writers," Hoebel (1949) explained, "have commented upon the relative absence of legislative enactment by primitive government" (p. 845). Often there was no authority competent to make a new rule: "It is seldom in the heads of a people to alter those customs which have been held sacred from time immemorial" (Westermarck, 1912, p. 162). This is because, as Lowie (1934) points out, the aim was "rather to exact obedience to traditional usage than to create new precedents" (p. 358). Indeed, as Sumner (1907, p. 355) explained, "The ghosts of the ancestors would be angry if the living should change the ancient folkways" (p. 355). Furer-Haimendorf (1967) claimed that Gond philosophy "leaves no doubt that the rules of behavior laid down in the ancestor's time remain binding for present generations" (p. 148).

The codes regulating interactions in traditional, kinship-based societies are said to focus on the roles of, and altruistic interactions between, kin. Four codes were said to be of fundamental importance because without these codes men would be "held down by low animal appetites and passions" (Morgan, 1877/1963, p. 41), return to a state of savagery, and live in misery (Tyler, 1881/1960). These consisted of codes which promoted motherhood (Edel & Edel, 1959), governed mate choice and marriage (Briffault, 1931; Coulanges, 1864/1955; Kroeber, 1948; Lowie, 1934; Malinowski, 1932; Rivers, 1998; Tyler, 1881/1960; Westermarck, 1912), encouraged respect for the elderly and the ancestors (Diamond, 1951; Santos Granero, 1991; Tyler, 1881/1960; Westermarck, 1912), and encouraged altruism toward kin (e.g., a male's offspring, one's siblings, *and far more distant kin*) (Edel & Edel, 1957; Tyler, 1881/1960; Westermarck, 1912). Given the overwhelming evidence that these traditional moral codes were crucial to the occurrence of much of the altruistic behavior that has characterized the social existence of our species, there is a need for an evolutionary explanation of these codes.

3.3 Parental Manipulation: Evolutionary Theory on the Verge of Understanding Traditions

Nearly 40 years ago, West-Eberhard summarized the evolutionary explanations of human altruism generated by the theoretical breakthroughs of Hamilton, Williams, Trivers, and Alexander: ". . . there are three general ways in which selection can act to produce beneficent social behavior: through kin selection, parental manipulation, and reciprocity" (West-Eberhard, 1975, p. 17). West, Mouden, and Gardner's (2011) recent review of the literature on the same subject demonstrated how kin selection and reciprocity have been widely used, and how group (or multilevel) selection has rebounded in popularity. However, West et al. (2011) make no mention of West-Eberhard's third evolutionary explanation of altruism. The recent relative neglect of the concept of parental manipulation is regrettable because evolutionary explanations of altruism based on parental manipulation stood on the verge of an explanation of the previously described traditional moral codes and the altruism they produced.

The parental manipulation explanation of altruism is based on the concept of parent-offspring conflict. As originally stated by Trivers (1974), the existence of parent-offspring conflict means that ". . . parents are expected to attempt to mold an offspring, against its better interests. . . ." (p. 249). This attempted molding, or manipulation, is the result of the simple biological fact that:

> The mother is equally related to [all of] her offspring. However, the offspring is completely related to itself [i.e., related to itself by 1.0], but only half as related to its full siblings [i.e., related to full siblings by 0.5]. A Hamiltonian offspring should value its personal fitness twice as much as it values any full sib's fitness. (Kurland & Gaulin, 2005, p. 452)

Therefore,

> Each child should, in theory, see itself as twice as valuable as its sibling [i.e., an offspring values itself 1.0 and values a full sibling 0.5], while the parent, being equally related to the two, values them equally. Hence another Darwinian prediction: not only will siblings have to be taught to share equally [i.e., taught to value a sibling as much as itself, or 1.0 instead of 0.5]; parents will, in fact, try to teach them [to value each sibling as much as itself, or 1.0]. (Wright, 1994, p. 166)

This generates the prediction that under certain circumstances, natural selection would favor parents who could *manipulate their offspring to behave as if each of the parent's other offspring were related to them by 1.0* (i.e., *value their siblings as much as they value themselves*). Although the chances of such total victory by a parent have long been the subject of debate (Alexander, 1974; Trivers, 1974), the outcome of parent-offspring conflict is likely to be some degree of compromise between the evolutionary interests of the parent and the offspring. It seems likely that parents who were more successful in this manipulation would sometimes be favored by natural selection over parents who were less successful. The power of this concept in explaining traditional moral codes and human altruism comes from the consequences of this parental manipulation when it is repeated in subsequent generations, a phenomenon made possible by human cultural traditions. Several theorists were tantalizingly close to recognizing the multigenerational consequences of parental manipulation during the 1970s and 1980s, but these consequences were never realized.

Trivers (1974) recognized that the influence of parents could extend far enough in time to alter "the later adult reproductive role of the offspring" (p. 262). Alexander (1974) elaborated on this point by stating that the tremendous potential for parental manipulation in humans is partially due to the long period in which living parents can manipulate their offspring's behavior: ". . . humans are parental manipulators par excellence. Their parental investment is enormous, and their generational overlap is extreme" (p. 367). Alexander (1974) also realized that this manipulation could continue even after the parent's death: ". . . humans may be unique among all organisms in that under normal circumstances a human offspring is never entirely without parental care, even if it has itself become a grandparent; even if its parents are dead, it will only rarely be without some direct benefits of parental care . . ." (p. 368). Alexander (1974) also appears to have realized that the longer the parent can manipulate the offspring to be altruistic, the better for the parent: "If individual offspring behave selfishly at termination of parental care . . . extensions of parental influence will be favored that encompass the detrimental situation, if they protect the brood from the selfish offspring or suppress the selfish behavior" (p. 345). Alexander (1974) even states that the ability of the parent to increase the altruistic behavior of the offspring does not necessarily have to end: "This multigenerational extension of parenthood has enormous significance in many regards . . . Since there is no obvious time at which parental care terminates . . ." (p. 368).

Two decades later Voland and Voland (1995) appear to be even closer to recognizing the full consequences of an infinitely extended parental manipulation when they propose that the human conscience is a means by which parents caused offspring to resist "selfish impulses" (1995, p. 401). They start by proposing that the existence of the human conscience, and the altruistic behavior it causes, is "not adequately explained by a mere reference to reciprocal altruism or kin selection" (p. 404) and that this represented a major gap in explanations of human altruism because the altruism produced by this extended form of parental manipulation ". . . finds its most remarkable expression in heroes and saints, but is by no means restricted to an ethical elite. It molds our daily life . . ." (1995, p. 404). Voland and Voland also recognized that conscience was one of the ways in which parental manipulation could continue to influence the behavior of offspring after the parent has died and it is no longer able to deter the selfish behavior of the offspring directly:

> The conscience evolved within the context of parent/offspring conflict over altruistic tendencies. As an extended phenotype of parental genes, it governs parental control on the offspring's behavior in a *lasting* way, even when there are no longer any direct possibilities for parental manipulation. (1995, p. 397; our emphasis)

Voland and Voland also write:

> The behavior of an individual should, therefore, not be hurriedly interpreted as being shaped by natural selection to the reproductive advantage of the gene programs of just this individual. The behavior being questioned can increase the genetic fitness of another individual as well. Consequently, organisms can serve replication interests with their behavior determined by others, either temporarily or *permanently*. (Voland & Voland, 1995, p. 404; our emphasis)

Perhaps the closest any statement came to realizing the full consequences of parental manipulation was: "They [offspring] were raised to 'voluntarily' stake at least part of their reproductive fitness for the maintenance and welfare of their families and thus to the long-term advantage of their *lineage*" (1995, p. 407; our emphasis). Unfortunately, instead of following this insight with the final step of realizing that lineages could be benefited through the transmission of a *tradition* of parental manipulation *indefinitely*, the authors return to only measuring the evolutionary success of parents: "The lifetime fitness of the altruist who is guided by his/her conscience and who acts ethically is negative, but not so for this altruist's manipulative parents, . . ." (Voland & Voland, 1995, p. 407). Thus, like the earlier theorists, Voland and Voland (1995) realized that parents can manipulate their offspring to engage in ". . . ethical and altruistic behavior—even long after the death of the parents" (p. 406), but they did not fully realize just how long after the death of a parent this production of ethical and altruistic behavior could continue.

All of the theorists just discussed appear to have recognized that parental behavior could produce altruism among offspring and grandchildren, but none of them appeared to have considered the possibility that parental manipulation could hypothetically produce altruistic behavior through an infinite number of generations of descendants if the manipulative behavior became traditional. What makes this situation all the more tantalizing is that several of the same theorists recognized the advantages, if not the necessity of, measuring evolutionary success over far more generations than are typically considered. None of them, however, linked this insight with the long-term influence of parental manipulation.

3.4 Measuring Evolutionary Success by the Effect of Traits on Future Generations

Alexander (1979) writes that it is crucial for evolutionary theorists to ask: ". . . What to measure, and what generation to measure it in, to determine which genetic line is winning (or what in fact constitutes "winning")" (p. 346). That is, "What is to be measured and when should it be measured? Should we measure numbers of offspring produced, numbers reared, numbers breeding, numbers of grandchildren produced, reared, breeding, *etc.*?" (Alexander, 1974, p. 374; emphasis added). The "*etc.*" is crucial because it indicates a realization that selection might be best measured as far into the future as possible, or at least further measurements would be superior to nearer ones when feasible. This implies that the success of a behavior might be different when measured further in the future than it is in the one or two generations where it is typically measured. West-Eberhard (1975) expands on this crucial point:

> This example raises the question of how far into future generations maternal control could be expected to operate. This raises the further general question of just what it is that selection maximizes—whether number of children, grandchildren, great-grandchildren or *n*th descendents (see Alexander, 1974)—and it shows another way in which classical fitness is an inadequate measure of an individual's total reproductive (genetic) contribution. Inclusive fitness can include effects on future generations but does not specify how many generations should be included. In threshold cases of hymenopteran sociality there must sometimes be a reduction in the mean fitness of the offspring—a paradox for classical theory. (Hamilton, personal communication) (p. 29)

Dawkins (1982) later made the same point even more clearly:

> Workers who correctly use the concept of fitness admit that it can be measured only as a crude approximation. If it is measured as the number of children born it neglects juvenile mortality and fails to account for parental care. If it is measured as number of offspring reaching reproductive age it neglects variation in reproductive success of the grown offspring.

If it is measured as number of grandchildren it neglects . . . And so on *ad infinitum*. Ideally we might count the relative number of descendants alive after some very large number of generations. (p. 184)

Unfortunately, the benefits, if not necessity, of measuring evolutionary success after many generations have not been pursued. For example, even when emphasizing the importance of a multigenerational approach, Lancaster and Kaplan (2009) merely refer to a "three-generational system of downward resource flow from grandparents to parents to children" (p. 95). In their previously cited review of the topic, West et al. (2011) feel comfortable in using a one generation measurement: "the quantity maximised by Darwinian individuals" is measured in the currency of "production of *offspring*" (p. 232; our emphasis). We will now explain how traditions extend the consequences of parental manipulation and then combine that with the benefits of measuring evolutionary success over many generations.

3.5 Ancestor-Descendant Conflict, Ancestor Manipulation, and Descendant-Leaving Success

Combining the ability of parental manipulation to cause altruism in descendants, the ability of human offspring to replicate the behaviors of their parents (i.e., traditions), and the measurement of evolutionary success over many generations (which we will refer to as descendant-leaving success in contrast to the one or two generational measurement of reproductive success) leads to the concept of "ancestor-descendant conflict" (Coe, Palmer, Palmer, & DeVito, 2010, p. 2). Parental manipulation became traditional when the parental manipulation of offspring that increased the altruism of the offspring toward each of the parent's descendants was *duplicated by those offspring and directed toward their own offspring* and so on and so forth through subsequent generations (see Steadman & Palmer, 1995). An oversimplified example of how parent-offspring conflict could have been transformed into ancestor–descendant conflict is the following three-part exhortation by a parent to his or her offspring: "1) treat all of my other descendants as if they are as valuable to you as you are to yourself, 2) tell your offspring to also treat all of my descendants as if they are as valuable to them as they are to themselves, and 3) also tell your offspring to tell their own offspring these things" (Coe et al., 2010, p. 6). A parent who started a *tradition* of parental manipulation of offspring to be more altruistic toward that parent's other descendants could increase that parent's number of descendants, and thus increase the numbers of copies of that parent's genes, in *distant future generations*.

Combining parental manipulation with tradition and the measurement of evolutionary success over many generations solves a puzzle in the cross-cultural record on kinship and altruism identified by Alexander. Alexander (1979) claims that for "most people in a modern technological society, . . . the significance of distinguishing relatives decreases beyond some level, such as that of first cousins, because of low relatedness. . . . [This is] obviously consistent with a Darwinian model" (pp. 148–149). The problem for current Darwinian explanations of human kinship is that the ethnographic data from traditional societies are obviously inconsistent with a Darwinian model because humans in every known traditional society are not only able to identify kin far beyond first cousins, but "extensive extra-familial nepotism" (Alexander, 1979, p. 211) also appears to be universal. Quoting Murdock (1949), Alexander describes this pattern by stating that universally:

> ...some of the intimacy characteristic of relationships within the nuclear family tends to flow outward along the ramifying channels of kinship ties . . . . [When an individual] needs assistance or services beyond what his family . . . can provide, he is more likely to turn to his secondary, tertiary, or remoter relatives than to persons who are not his kinsmen. (As referenced by Alexander, 1979, p. 156, from Murdock, 1949)

In a recent overview of evolutionary explanations of kinship altruism, Bernstein (2005) elaborates on exactly why this feature of human kinship found in traditional societies is "surprising" to an evolutionist:

> Because the return to fitness of altruism toward distant distant kin [i.e., kin far more distantly related than first cousins] is miniscule, typically less than helping an unrelated person with whom another has a profitable exchange, it may be surprising that such groups often have norms obliging members to favor these distant distant relatives over non-kin . . . If altruism is prescriptive even on occasions when the degree of genetic relatedness is very small, the altruist's fitness will decline depending on the frequency of such occasions. (p. 529)

Bernstein (2005) suggests that such puzzling altruism can be ignored because it primarily occurs in situations such as famine and war. Even if this is true, and it is clearly debatable, it still begs the question of why such altruism, and the norms obliging individuals to engage in such altruism, should occur at all. It also begs the question of why Alexander pointed out that such an extension of altruism appears to be universally found in traditional societies, but not in nontraditional "modern" societies. Thus, mainstream evolutionary explanations of altruism are left with the puzzle that the concept of ancestor–descendant conflict solves.

We now turn to a brief description of how the disruption of traditions of parental manipulation that have led to the reduced scope of kinship altruism in nontraditional societies may have started in early state societies that attempted to incorporate multiple kinship-based societies.

3.6 Changes from Moral Codes Toward Kin to Law Toward Non-kin in Early States

Although often appealing to earlier aspects of traditional moral codes for legitimacy, the earliest laws associated with the emergence of the commonwealth (e.g., Mosaic law, Hammurabi's codes) were not themselves traditional. Instead their *source* was a *new* supernatural revelation. The laws of Moses were said to have come to him through divine revelation from the ancestor, Yahweh, who created them; those of Hammurabi of Babylon were said to have been revealed from the Sun-God Samas, the judge of Heaven and Earth (Johns, 1903). Mosaic laws were said to have been revealed to the prophet Moses in order to regulate the behavior of a group of individuals who were "not community of blood, or of land, or of government...but a crowd of mixed ancestry which fled Egypt" (Suelzer, 1964, p. 90). Laws also are said to be "enhanced by the belief that they are fair and just" (Schwartz & Rosenbaum, 1983, p. 241). However, this may be mere rhetoric. It seems clear that "equity is not a necessary condition for the constitution of law; even a shockingly unjust decision...can be law" (Van Baal, 1981, p. 111).

In early nation states the *scope* of laws is geographic, including the entire nation state, and thus includes non-kin. This shift makes the concept of "a group" more plausible. Schapera (1956) explained that a state or commonwealth is not a closed group with membership determined solely and permanently by descent. It is rather an association into which people may be born, absorbed by conquest, or admitted as immigrants and from which they may depart voluntarily or be driven by the fortunes of war. Hammurabi's code, for example, brought together in one geographic area two unrelated groups, Sumerians and Semitics (Diamond, 1951). The foundation of Israel, according to Suelzer (1964), was not "community of blood or land, or of government," it was "alliance with the lord [which] united the crowd of mixed ancestry which fled Egypt" (p. 90). In other words, early law created metaphorical kinship ties among non-kin, united by a prophet, Moses, who spoke for an ancestor, Yahweh, who was the father of all men. The boundaries of the Promised Land were said to have been established by God (Deut iv 6). Tribes living outside the geographical area and not sharing the Hebrew God were neither protected by nor subject to Mosaic law (Wines, 1853).

In kinship-based moral system, the codes focus on altruism toward kin and come from the ancestors of those kin, and obeying them is a duty owed to one's ancestors. The codes specify altruistic kinship behavior and the most serious offenses against the ancestors, and his/her descendants, are exile, or the loss of all kinship ties. The source of the codes in a legal system is also said to be an ancestor, actual or metaphorical. The ability to influence behavior, in both systems, whether of an elder or a leader, depends upon ancestral endorsement and leadership in both systems and is defined more by obligations than privileges. Leadership is legitimate when it has both an ancestral endorsement and evidence of responsiveness to followers (co-descendants) and fulfillment of obligations to them.

In a moral system, there is no system for the creation of new codes, as the codes themselves are largely immutable. Legal systems, however, have methods and mechanisms in place for legislative enactment. Although legal systems also have immutable codes (which are said to be ancestral and which focus especially on such things as honoring the elders), a new type of code has emerged. These codes, which are mutable, focus on temporary relationships between buyer and seller.

The system found in the early state differs from the one found in kinship-based, traditional societies primarily in the degree of formality because, for example, the laws have been written down. There is also a difference in the power of authority. To some extent, the education of children, or transmission of knowledge about the system and its rules, is not accomplished by modeling, storytelling, or other informal methods but rather has been taken out of the hands of parents and placed in the hands of the state. In order to finance the system, tributes are specified and must be paid on a regular basis. Tributes are no longer made, as sacrifices, to the ancestors but are given to the ancestor's living representative. In this system, attention is paid to the punishments specified for particular offenses (e.g., the punishment for treason is death), and less credence is given to contingencies that may have influenced why one committed an offense.

3.7 Conclusion

Thomas Hobbes (1651) exemplifies the common view that before there were legal social contracts, there was "war of all against all." The cross-cultural study of traditional pre-state societies, however, leads to a very different conclusion. Before the legal systems of early states, there were traditional moral systems aimed at promoting altruism among co-descendants, including vast numbers of very distantly related co-descendants. While legal codes are aimed at regulating selfishly motivated interactions among non-kin, moral codes are aimed at promoting the well-being of descendants. Evolutionary psychology may currently have no theory that allows us to explain the "axiom of kinship amity" or the cooperative treatment of those identified by descent from a common ancestor. As the behavior is so widespread, however, now may be the time when hypotheses, such as the one based on cultural traditions, are proposed and tested against the cross-cultural evidence.

References

Alexander, R. (1974). The evolution of social behavior. *Annual Review of Ecology and Systematics, 5*, 325–383.

Alexander, R. (1979). *Darwinism and human affairs*. Seattle, WA: University of Washington Press.

Bandalier, G. (1972). *Anthropological politique*. Hammondworth: Penguin.

Bernstein, F. (2005). Altruism and genetic relatedness. In D. M. Buss (Ed.), *The handbook of evolutionary psychology* (pp. 528–551). Hoboken, NJ: John Wiley and Sons.

Briffault, R. (1931). *The mothers: The matriarchal theory of social origins*. New York: Macmillan.

Campbell, D. T. (1975). On the conflicts between biological and social evolution and between psychology and moral tradition. *American Psychologist, 30*, 1103–1126.

Coe, K. (1995). *Messages from the ancestors: Moral systems and laws of the Chachi of Lowland Ecuador*. Dissertation, Arizona State University.

Coe, K. (2003). *The ancestress hypothesis*. Newark, NJ: Rutgers University Press.

Coe, K., Palmer, A. L., Palmer, C. T., & DeVito, C. L. (2010). Culture, altruism, and conflict between ancestors and descendants. *Structure and Dynamics, 4*(3), 1–17.

Coulanges, F. (1955, first published in 1864). *The ancient city*. Garden City, NY: Doubleday and Company.

Cronk, L. (1999). *That complex whole*. Boulder, CO: Westview Press.

Culwick, A. T., & Culwick, G. M. (1935). *Ubena of the rivers*. London: G. Allen and Unwin, Ltd.

Dawkins, R. (1982). *The extended phenotype*. Oxford: Oxford University Press.
Diamond, A. (1951). *The evolution of law and order*. Westport, CT: Greenwood Press.
Edel, M., & Edel, A. (1957). *Anthropology and ethics*. Springfield, IL: Charles C. Thomas.
Evans-Pritchard, E. E. (1951). *The Nuer: A description of the modes of livelihood and political institutions of a Nilotic people*. Oxford: Oxford University Press.
von Furer-Haimendorf, C. (1967). *Tribal populations and cultures of the Indian subcontinent*. Leiden, NL: E. J. Brill.
Hobbes, T. (1651) Pirated edition of *The elements of law, natural and politic*, repackaged to include two parts. *Human nature, or the fundamental elements of Policie* and *De Corpore Politico*. Retrieved March 22, 2013 from http://www.constitution.org/th/elements.htm.
Hoebel, E. A. (1949). *Man in the primitive world: An introduction to anthropology*. New York: McGraw-Hill.
Johns, C. (1903). *The oldest code of law in the world*. Edinburgh, SCT: T and T. Clark.
Johnson, O. (1984). *Ethics: Selections from classical and contemporary writers* (5th ed.). New York: Hold, Rinehart, and Winston.
Keesing, R. M. (1975). *Kin groups and social structure*. New York: Holt, Rinehart and Winston.
King, P. (1972). *Law and society in the Visigothic kingdom*. New York: Vantage Press.
Kroeber, A. L. (1948, first published in 1923). *Anthropology* (Revised ed). New York: Harcourt, Brace.
Kurland, J. A., & Gaulin, S. (2005). Cooperation and conflict among kin. In D. M. Buss (Ed.), *The handbook of evolutionary psychology* (pp. 447–482). Hoboken, NJ: John Wiley and Sons.
Lancaster, J. B., & Kaplan, H. S. (2009). The endocrinology of the human adaptive complex. In P. T. Ellison & P. B. Gray (Eds.), *Endocrinology of social relationships* (pp. 95–118). Cambridge, MA: Harvard University Press.
Lowie, R. (1934, first published in 1919). *Primitive society*. New York: Ferrar and Reinhart.
Malinowski, B. (1932). *The sexual life of savages* (3rd ed.). London: Routledge.
Middleton, J. (1960). *Lugbara religion: Ritual and authority among an East African people*. London: Oxford University Press.
Morgan, L. H. (1963, first published in 1877). Ancient society. Cleveland, OH: The World Publishing Company.
Murdock, G. P. (1949). *Social structure*. New York: MacMillan.
Palmer, C. T. (2010). Cultural traditions and the evolutionary advantages of non-innovation. In M. O'Brien & S. Shennan (Eds.), *Innovation in cultural systems: Contributions from evolutionary anthropology* (pp. 161–174). Cambridge, MA: MIT Press.
Palmer, C. T., & Steadman, L. (1997). Human kinship as a descendant leaving strategy: A solution to an evolutionary puzzle. *Journal of Social and Evolutionary Systems, 20*(1), 39–51.
Rattray, R. S. (1929). *Ashanti law and constitution*. Oxford, England: Clarendon Press.
Rivers, W. H. R. (1998, first published in 1917). *Kinship and social organization*. London: Rutledge/Thoemmes Press.
Salter, F. (2002). *Risky transactions: Trust, kinship, and ethnicity*. New York: Berghahn.
Santos Granero, F. (1991). *The power of love: The moral use of knowledge amongst the Amuesha of Central Peru* (London school of economics monographs on social anthropology, Vol. 62). London: The Athlone Press.
Schapera, A. (1956). *Government and politics in tribal societies*. London: Watts.
Schwartz, C., & Rosenbaum, T. (1983). Allometric influences on primate mothers and infants. In L. Rosenbaum & H. Moltz (Eds.), *Symbiosis in parent-offspring relationships*. New York: Plenium.
Steadman, L. B., & Palmer, C. T. (1994). Visiting dead ancestors: Shamans as interpreters of religious traditions. *Zygon: Journal of Religion and Science, 29*(2), 173–189.
Steadman, L., & Palmer, C. T. (1995). Religion as an identifiable traditional behavior subject to natural selection. *Journal of Social and Evolutionary Systems, 18*(2), 149–164.
Steadman, L., Palmer, C. T., & Tilley, C. F. (1996). The universality of ancestor worship. *Ethnology, 35*(1), 63–76.
Suelzer, A. (1964). *The pentateuch*. New York: Herder and Herder.

Sumner, W. (1907). *Folkways: A study of the sociological importance of usages: Manners, customs, mores and morals*. Boston, MA: Genn and Company.
Sun Chief. (1942). *The autobiography of a Hopi Indian*. New Haven, CT: Yale University Press.
Trivers, R. L. (1974). Parent-offspring conflict. *American Zoologist, 14*, 249–264.
Turner, V. (1979). Divination as a phase in a social process. In W. Lessa & E. Vogt (Eds.), *Reader in comparative religion: An anthropological perspective* (4th ed., pp. 373–376). New York: Harper Collins.
Tyler, E. (1960, first published in 1881). *Anthropology*. Ann Arbor, MI: University of Michigan Press.
Van Baal, J. (1981). *Man's quest for partnership*. Assen, The Netherlands: Van Gorcium.
van den Berghe, P. (1979). *Human family systems: An evolutionary view*. New York: Elsevier.
van den Berghe, P., & Barash, D. (1977). Inclusive fitness and human family structure. *American Anthropologist, 79*, 809–823.
Voland, E., & Voland, R. (1995). Parent-offspring conflict, the extended phenotype and the evolution of conscience. *Journal of Social and Evolutionary Systems, 18*, 387–412.
West, S. A., Mouden, C., & Gardner, A. (2011). Sixteen common misconceptions about the evolution of cooperation in humans. *Evolution and Human Behavior, 32*, 231–262.
West-Eberhard, M. J. (1975). The evolution of social behavior by kin selection. *The Quarterly Review of Biology, 50*(1), 1–33.
Westermarck, E. (1912). *The origin and development of the moral ideas* (Vol. 1). London: Macmillan.
Wines, E. (1853). *Commentaries on the laws of the ancient Hebrews*. New York: Putnam.
Wright, R. (1994). *The moral animal*. New York: Vintage Books.

Chapter 4
Guatemalan Adolescents' Reports of Helping in Urban and Rural Mayan Communities

Judith L. Gibbons

4.1 Introduction

Altruism, "a motivational state with the ultimate goal of increasing another's welfare" (Batson & Shaw, 1991, p. 6), has been the focus of a great deal of research in recent years. It is evident that altruistic behaviors are widespread throughout human societies and in nonhuman primates. Researchers have shown that even young children may act altruistically (Svetlova, Nichols, & Brownell, 2010). By 12 months of age, children may comfort someone in distress or retrieve out-of-reach objects. By 2 years of age, they deliver food to both themselves and another person, rather than to themselves alone (Brownell, Svetlova, & Nichols, 2009). The early emergence of helping behavior in children, as well as its appearance in other species, has led social evolutionary theorists to posit that altruism has evolved in humans because it is adaptive for the gene pool (see review by Bshary & Bergmüller, 2008). Despite a genetic contribution to helping behavior, the major role of culture in the expression of altruism has been well established (see Chap. 3 of this volume by Coe and Palmer (2013)).

Moreover, neuroimaging studies have provided evidence for the involvement of particular neural systems in empathy and altruistic motivation (e.g., Cheon et al., 2011; Huffmeijer, Alink, Tops, Bakermans-Kranenburg, & van IJzendoorn, 2012; Mathur, Harada, Lipke, & Chaio, 2010). The particular brain regions involved may differ depending on whether the altruism is expressed toward members of the ingroup or the outgroup (Cheon et al., 2011; Mathur et al., 2010). On the other hand, the hormone oxytocin may promote altruism toward both ingroup and outgroup members, as shown in studies in which the hormone was experimentally

J.L. Gibbons (✉)
Department of Psychology, Saint Louis University and Universidad del Valle de Guatemala, 221 North Grand Blvd, St. Louis, MO 63103, USA
e-mail: gibbonsjl@slu.edu

D.A. Vakoch (ed.), *Altruism in Cross-Cultural Perspective*, International and Cultural Psychology, DOI 10.1007/978-1-4614-6952-0_4, administered through nasal spray (Barraza, McCullough, Ahmadi, & Zak, 2011; Israel, Weisel, Ebstein, & Bornstein, 2012). Because of the overlap in the neural and hormonal systems involved in altruism and parenting, researchers have suggested that a caregiving system has evolved in humans that promotes nurturance of children and helpfulness toward others (Swain et al., 2012).

At the level of the individual, altruistic behaviors may vary depending on personality and status. A recent series of studies showed that humble persons were more helpful than non-humble ones (LaBouff, Rowatt, Johnson, Tsang, & Willerton, 2012). In general, people with fewer resources and lower status are more compassionate, altruistic, and generous than others (Kraus, Piff, Mendoza-Denton, Rheinschmidt, & Keltner, 2012; Piff, Kraus, Côté, Cheng, & Keltner, 2010; Rucker, Dubois, & Galinsky, 2011; Stellar, Manzo, Kraus, & Keltner, 2012; Visser & Roelofs, 2011). Also, people who have suffered may be more altruistic, especially toward disadvantaged outgroups (Vollhardt & Staub, 2011).

Most cross-cultural studies of altruism have focused on the demographic characteristics of communities that predict helpfulness toward strangers (e.g., Levine, Norenzayan, & Philbrick, 2001; Levine, Reysen, & Ganz, 2008). In those studies, confederates posed as a pedestrian dropping a pen, a person with an injured leg trying to recover dropped magazines, and a blind person attempting to cross the street. Among cities in the United States, residents of cities with high population density, economic purchasing power, and fast walking speed were less likely to help a stranger (Levine et al., 2008). In an international study of 23 large cities, the same methodology revealed large differences in helping that were significantly related (negatively) to economic prosperity (Levine et al., 2001); that is, less helping behavior was demonstrated in wealthier cities. Cultural individualism-collectivism did not correlate with helpfulness across cities.

The literature on altruism sheds little light on altruism under naturalistic conditions, helping in developing countries or helping kin rather than strangers. The last is particularly important, because a meta-analysis has revealed kinship as the strongest predictor of altruism (Kruger, 2003). Overall, there is little information from the everyday contexts in which altruism occurs, especially in the developing world.

4.2 Altruism During Adolescence

The life stage of adolescence may be particularly important for the development and maintenance of altruism. Developmentally, an increase in cognitive abilities at adolescence allows teenagers to engage in more elaborate and mutual perspective-taking (Selman, 1980); better perspective-taking skills have been linked empirically to increased helping behavior (Eisenberg, Cumberland, Guthrie, Murphy, & Shepard, 2005). In addition, an international study of over 8,000 adolescents revealed that kindness, along with honesty, was the quality that was most highly valued in the ideal person (Gibbons & Stiles, 2004). In that study, approximately one-fourth of the adolescents spontaneously drew the ideal person helping or caring for others. Correlational studies have shown that engaging in volunteer activities or fund-raising as an adolescent is related to altruism during adulthood (Rosen & Sims, 2011). The authors suggest that altruism may be habit-forming. In another correlational study, engaging in helping behavior was associated with positive social relations and greater well-being in adolescents, although there were some gender differences in the relations (Schwartz, Keyl, Marcum, & Bode, 2009). A study conducted in 13 countries revealed that students were more likely to volunteer than were adults and that the motives they gave for volunteering were most often altruistic (see Chap. 6 of this volume by Grönlund (2013)). Finally, interventions that promote volunteerism in adolescents lead to positive outcomes for adolescents. For example, in an experimental study of the Teen Outreach Program, adolescents randomly assigned to the intervention that involved volunteering showed more academic success and fewer problems such as early pregnancy and school dropout (Allen, Philliber, Herrling, & Kuperminc, 1997). In sum, adolescence may be a critical period for the development of altruistic motivation and helpful behavior, and altruism may be beneficial to adolescent well-being.

4.3 Child Development and Collaboration in Indigenous Communities

A marked feature of child development in indigenous communities of the Americas is the emphasis on child collaboration and participation in the daily activities of the household (López, Najafi, Rogoff, & Mejía-Arauz, 2012).

In Yucatec Mayan communities in Mexico, children from an early age are routinely involved in everyday household tasks (Gaskins, 2008). From ages 3–6, children often independently bathe and dress themselves or do simple tasks, although they also spend a great deal of time observing the work of adults and older siblings. Beginning at age 6, they may take care of younger siblings, run errands, or help older siblings with chores outside the house. By age 12 and beyond, about 60 % of adolescents' time is spent doing significant chores that benefit the family. For example, in a single day young adolescent Pancho weeded the family cornfield all morning with his father. During the afternoon he fed the pigs, went shopping, and helped his father repair a fence. His sister Juana helped her mother prepare food, clean the house, and do laundry; she also went shopping. According to Gaskins (2003), children's participation in and observation of daily activities prepare them for future roles in the community.

In a Mayan community in Guatemala, Rogoff and her colleagues observed that children learned primarily through "guided participation" in cultural activities (Rogoff, Mistry, Göncü, & Mosier, 1993). In a subsequent study, Mayan mothers with little education engaged children in "horizontal multiparty engagements." In other words, children and mothers worked collaboratively on tasks, without mothers taking charge and supervising (Chavajay & Rogoff, 2002).

In summary, children in indigenous communities are often incorporated into ongoing activities. They participate eagerly and may receive guidance from others who are present (López et al., 2012). A cultural value that may be associated with children's collaboration in indigenous communities is represented by the Spanish word *acomedido/a* (López et al., 2012). It connotes helpfulness and accommodation to ongoing activity. An acquaintance of mine, a young father, described how his 17-month-old son carries pieces of firewood to the woodpile and tries to sweep the floor, "*Es muy inteligente* [he is very intelligent]," he said proudly.

At a societal level, Mayan communities are well known for their cargo system, a practice that redistributes wealth (Cancian, 1965). Under the cargo system, families that have achieved greater economic success assume religious and civic responsibilities that mandate, not only service, but also economic contributions to the welfare of the community. Participation, while financially costly, increases the status of the family in the community. For a thorough discussion of cargo as altruism in Mesoamerica, see Chap. 11 of this volume by Chick (2013). The expectation that children will contribute to the family well-being may be early training for participation in community activities such as cargo.

4.4 Photographic Methods in Research

> As researchers, we see ordinary people's everyday lives as interesting, complex, and having the potential to reveal important understandings about the human condition. (Rich, 2002, p. 409)

A valuable tool in understanding ordinary people's everyday lives is the method known as photovoice, or auto-photography (Wang & Burris, 1997). In this procedure, individuals take photographs that reveal their lives or an aspect of their lives. Because the participants control the content of their depictions and actively engage in telling their stories, photovoice has been seen as a means to empower individuals (Castleden, Garvin, & Huu-ay-aht First Nation, 2008). Photovoice may be a particularly valuable procedure for adolescents from indigenous communities who may have limited literacy and may prefer to express themselves visually rather than in words. Among rural home economics students in the United States, Blinn and Schwartz (1988) found that adolescents' photographs of their future depicted occupations, families, home ownership, and driving/automobiles. In another study that used photovoice with adolescents, students took photographs of what made them feel safe, happy, and healthy, or unsafe, unhappy, and unhealthy. From their photographs, they developed group social action projects to address those issues (Wilson et al., 2007).

4.5 The Present Study

The present study was designed to understand the lives of Guatemalan adolescents living in distinct communities—a rural indigenous K'iche'-speaking community and an urban community. A fuller report of the findings can be found in Gibbons (2003). Based on previous literature, it was expected that adolescents from the rural indigenous communities would depict more images and more often mention helping others than would adolescents from the urban setting.

In this rural community of several thousand persons, the economy was based on agriculture; the primary crops were corn, apples, and plums. The community had been influenced by national and local changes. The Guatemalan Peace Accords of December 1996 had mandated bilingual education in Mayan communities. At the time of the study, most teachers were fluent in the local Mayan language, K'iche' (as well as Spanish), although few could read and write it. At the time of the study, the recent completion of a road (two concrete paths) had made the village more accessible.

For comparison, the urban setting was in Antigua, Guatemala, a colonial town of about 25,000 people known for its colonial architecture and as a tourist destination. The economy is based on tourism and coffee production in the surrounding areas; the local market attracts both Guatemalans from surrounding towns and tourists.

The participants were 73 adolescents, ages 11–16 (32 boys, 41 girls; 30 from the urban school, 43 from the rural school). Students from the public rural school were in grades 4–6. Urban participants attended a private school, known for its strict discipline; the tuition of about \$13 per month is well beyond the means of many Guatemalans. Urban students were in *básico* (junior high school), in grades 7–9. Thus, although ages were matched, the urban sample had experienced more years of education.

In the year 2000, with the approval of the Institutional Review Board, school officials, and parents, 73 adolescents were provided with disposable cameras (with flash) through their school classrooms. Based on the procedure of Blinn and Schwartz (1988), their instructions were to take 10 photographs that represented their life in the present. For each photograph, answer the following questions: (1) Who took the picture? (Use no names. Write myself, my friend, my sister, etc.) (2) What is the picture of? (3) Where was the picture taken? (4) What aspect of your life does the photo show? (5) How do you feel about this? Then take 10 photographs that represent your life in the future. For each photograph, answer the following questions: (1) Who took the picture? (2) Where was the picture taken? (3) What aspect of your life does the photo show? (4) How do you feel about this? (5) How old will you be when this happens? Take the remaining seven photos of anything you would like. In addition, several demographic questions, such as age and year in school, were included. The cameras and instruction sheets were distributed and photography explained and demonstrated. Approximately 2 weeks later, the cameras and questionnaires were collected from the participants. After the photographs were developed and printed, copies of the photographs were returned to each participant during a brief interview in which questionnaire responses were matched to the photographs.

Categories were developed for scoring the photographs, and each photograph was scored by three raters. The reliability of coding as determined by intraclass reliability coefficients ranged from .57 for religious symbols to .98 for number of males in the picture. The responses to questionnaires on the questionnaire were typed (maintaining grammatical and spelling errors) and then translated from Spanish into English (with spelling and grammar corrected.) The verbal samples were then analyzed in 74 categories using the LIWC program (Pennebaker, Francis, & Booth, 2001). In addition, the verbal samples were scored for mentioning help or helping (*ayuda, ayudar*) in three categories: helping family members, helping people outside the family, and being helped. The total number of comments about help was also recorded.

4.6 Results

Participants reported that they enjoyed the exercise; many said that they had never before used a camera. In addition, they said that it was an interesting way to learn something new.

Overall, 33 % of the participants mentioned helping in their comments. See Table 4.1 for the types of helping mentioned and sample statements.

The number of comments in each category and the mean number of help comments overall were analyzed by a 2 Gender by 2 Community analysis of variance (ANOVA).[1] The analysis revealed that, contrary to predictions, students in the urban community mentioned help (a mean of 1.6 times total mentions) more than did

Table 4.1 Prevalence of comments about help

Category of comment	Percentage of participants	Examples
Help family members	9.6	Helping my mom
		I have to help my brother
Help others, not family	24.7	Help my fellow man
		I would like a project in the future to help children
Help me or us	14.1	This helps us. It helps me to see how I am
Any mention of help	32.9	Any of the above

[1] Analysis of variance (ANOVA) is a statistical technique for determining whether the mean differences between groups are likely to have occurred by chance. A p value of $< .05$ is usually the criterion for significance and indicates that the probability is less than one in 20 that the difference occurred by chance. η^2 (eta-squared) is the effect size, the proportion of the total variance accounted for by the evaluated effect.

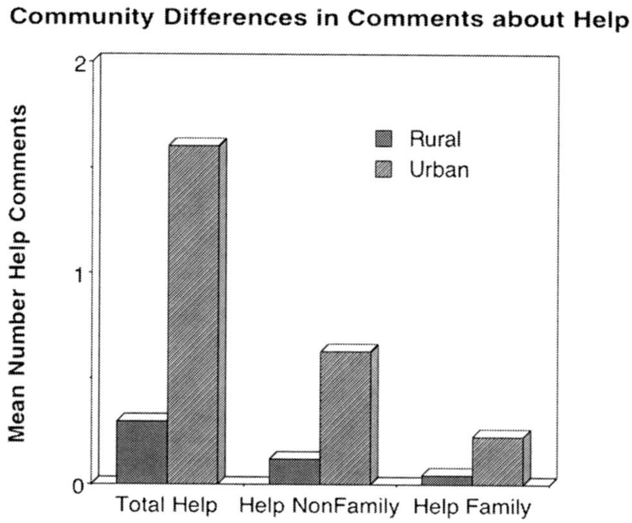

Fig. 4.1 Mean number of help comments in the two communities

students in the rural community (a mean of .32 times total mentions), F (1, 69) = 10.41, p = .002, and η^2 = +.105. There was no significant effect for gender. For mentioning helping others, not family members, there was a significant effect for community, F (1, 69) = 12.63, p = .01, and η^2 = .113, with urban adolescents more likely to mention helping those outside the family. Boys mentioned helping others outside the family more than did girls (mean for boys = .55, mean for girls = .22), F (1, 69) = 5.64, p = .02, and η^2 = .050. There was a trend toward adolescents in the urban community mentioning helping family members more than adolescents in the rural community, F (1, 69) = 3.84, p = .054, and η^2 = .141. Boys reported helping family members more than did girls, F (1, 69) = 10.76, p = .002, and η^2 = .114. There were no significant community or gender differences for mentioning being helped. These results are graphically presented in Fig. 4.1.

A typical photograph is depicted in Fig. 4.2. This photo was taken by a 16-year-old boy from the rural community. He wrote that, "my Dad [took the picture]," "of the bakery," "in the bakery," "it is good to help my Dad," "happy, because making bread is very good." The face has been blurred to maintain anonymity.

The photographs were also scored for the total number of children present (in all 20 photos). A 2 Gender by 2 Community ANOVA revealed that girls took more photographs of children than did boys (girls' mean = 9.03, boys' mean = 4.28), F (1, 66) = 5.06, p = .028, and η^2 = .034. In addition, adolescents from the rural community took more photographs of children than did adolescents from the urban community (rural adolescents' mean = 10.31, urban adolescents' mean = 3.00), F (1, 66) = 11.93, p = .001, and η^2 = .080.

Based on data from the LIWC program, the frequency of using the word "we" was also analyzed by a 2 Gender by 2 Community ANOVA. The results showed that

Fig. 4.2 A photograph taken by a 16-year-old boy from the rural community

the word "we" was used more by adolescents from the rural community than by those from the urban community (mean for rural adolescents = 2.93, mean for urban adolescents = .88), F (1, 68) = 9.58, p = .003, and η^2 = .074. Gender differences were not significant.

4.7 Discussion

Although children and adolescents in indigenous Mayan communities participate extensively in family chores (López et al., 2012), this study did not confirm the hypothesis that adolescents from the rural Mayan community would report more altruism, defined as making comments about "helping" in describing their lives. In the present study, Guatemalan adolescents who attended a private school in an urban community mentioned helping others, both family members and persons outside the family, more often than did the rural adolescents. This finding at first glance not only refutes the hypothesis but also appears to be inconsistent with the finding that persons with fewer resources are more altruistic than persons with more resources or higher status (Kraus et al., 2012).

Another possible interpretation of these results is that the rural adolescents took their participation in family chores for granted and did not perceive their cooperation as "helping" but simply as their own role within the family. To encode a behavior as helping, one needs first to see the activity as another person's responsibility. It is possible that in indigenous families, adolescents see their participation in productive activities as their own responsibility and not as a favor to someone else.

A recent study of cooperative hunting among Martu women revealed that skilled hunters incurred costs; solitary hunting yielded more food (Bird, Scelza, Bird, & Smith, 2012). One of the better hunters explained her altruistic behavior:

> I like to hunt with someone who is going to share. If we hunt together, I will give you half of what I get. Then you have something to share even if you didn't get lucky yourself. You're looking out (*kanyinin*) for someone else. (p. 75)

Although the hunter did see her behavior as cooperative, it was not explicitly helping another person.

In the present study, some indirect evidence comes from the additional analyses of the numbers of children in the photographs of the participants and in the use of the word "we." One of the primary tasks of adolescent girls in preindustrial societies is the care of younger siblings (Rogoff et al., 1993; Schlegel & Barry, 1991). The greater presence of younger children in the photos of rural adolescents, specifically those taken by girls, suggests that the girls may be caring for younger siblings (without coding the activity as "helping" their mothers). The more frequent use of the word "we" among rural adolescents also suggests that, rather than thinking of an "I" who helps others, they might have been thinking of the tasks that "we" do as a community or family.

In the present study, helpfulness was also mentioned more often with respect to others outside the family than toward family members. This finding also contradicts the established pattern that altruism is more often expressed toward kin than toward non-kin (e.g., Kruger, 2003).

Overall, the use of verbal expressions of "helping" has revealed unexpected findings that can best be understood as participants expressing the noteworthy, rather than the taken-for-granted, aspects of their lives. Adolescents from the rural community did not describe their contributions to the ongoing tasks as "helping"; contributing to the family well-being was part of their expected role.

According to some researchers, behavior that is undertaken as part of one's social role is not altruism (see Chap. 2 of this volume by Smith et al. (2013)). Thus, according to the definition put forth by Smith, Smith, Pieper, Yoo, Ferris, Downs, Bowden, and Butler (2006), the urban adolescents' contributions to the family, which are not part of their expected role, are altruistic, while similar contributions by rural adolescents are not. Similarly, gender roles prescribe that girls help care for younger siblings; boys performing the same behaviors would be acting altruistically. In their use of the word "help" to describe their activities, adolescents in the present study appear to echo the definition of Smith et al. (2006). However, cultural and gender differences in social roles make this conception of altruism complex and variable.

Although the photovoice method yields a rich, thick description[2] of adolescents' perceptions of their lives, their perceptions may not be closely correlated with their actual behaviors. The next step in studying altruism among Guatemalan adolescents

[2] Clifford Geertz (1973) in *The Interpretation of Cultures* used the term "thick description" to mean a detailed contextualized account that allows the ethnographer to understand the meaning of behavior.

would be to heed the advice of Pawlik (2012) to evaluate behavior in its natural ecology. The technique of "ambulatory assessment," involving following participants and noting their behavior as it occurs, would be particularly useful. Independent of whether they encode their behavior as helping others, do Guatemalan adolescents engage in helping and other altruistic behaviors as they go about their daily lives?

The study of altruism where it occurs in everyday life is still in its infancy. To understand how altruism is enacted within societies, research that uses multi-methods in a variety of cultures and settings is necessary.

Acknowledgments The author would like to thank Samantha Wilson, Natalie Humphrey, Sandy Smith, and Christine Rufener for sorting and coding photographs and entering data.

References

Allen, J. P., Philliber, S., Herrling, S., & Kuperminc, G. P. (1997). Preventing teen pregnancy and academic failure: Experimental evaluation of a developmentally based approach. *Child Development, 64*, 729–742.

Allen, J. P., & Philliber, S. (2001). Who benefits most from a broadly targeted prevention program?: Differential efficacy across populations in the teen outreach program. *Journal of Community Psychology, 29*, 637–655.

Barazza, J. A., McCullough, M. E., Ahmadi, S., & Zak, P. J. (2011). Oxytocin infusion increases charitable donations regardless of monetary resources. *Hormones and Behavior, 60*, 148–151.

Batson, C. D., & Shaw, L. L. (1991). Evidence for altruism: Pluralism of prosocial motives. *Psychological Inquiry, 2*, 107–122.

Bird, R. B., Scelza, B., Bird, D. W., & Smith, E. A. (2012). The hierarchy of virtue: mutualism, altruism and signaling in women's cooperative hunting. *Evolution and Human Behavior, 33*, 64–78.

Blinn, L. M., & Schwartz, M. (1988). Future time perspective: A multi-method study of how home economics students picture their lives in the future. *Journal of Vocational Home Economics Education, 6*, 1–17.

Brownell, C. A., Svetlova, M., & Nichols, S. (2009). To share or not to share: When do toddlers respond to another's needs? *Infancy, 14*, 117–130.

Bshary, R., & Bergmüller, R. (2008). Distinguishing four fundamental approaches to the evolution of helping. *Journal of Evolutionary Biology, 21*, 405–420.

Cancian, F. (1965). *Economics and prestige in a Maya community: The religious cargo system in Zinacantan*. Stanford, CA: Stanford University Press.

Castleden, H., Garvin, T., & Huu-ay-aht First Nation. (2008). Modifying photovoice for community-based participatory indigenous research. *Social Science & Medicine, 66*, 1393–1405.

Chavajay, P., & Rogoff, B. (2002). Schooling and traditional collaborative social organization of problem solving by Mayan mothers and children. *Developmental Psychology, 38*, 55–66.

Cheon, B. K., Im, D., Harada, T., Kim, J-S., Mathur, V. A., Scimeca, J. M.,... Chiao, J. Y. (2011). Cultural influences on neural basis of intergroup empathy. *NeuroImage, 57*, 642–650.

Chick, G. (2013). Altruism in human ritual. In D. A. Vakoch (Ed.), *Altruism in cross-cultural perspective*. New York: Springer.

Coe, M. K., & Palmer, C. T. (2013). Cross-cultural variation in altruism: Traditional parental manipulation and ancestor–descendant conflict. In D. A. Vakoch (Ed.), *Altruism in cross-cultural perspective*. New York: Springer.

Eisenberg, N., Cumberland, A., Guthrie, I. K., Murphy, B. C., & Shepard, S. A. (2005). Age changes in prosocial responding and moral reasoning in adolescence and early adulthood. *Journal of Research on Adolescence, 15*, 235–260.

Gaskins, S. (2003). From corn to cash: Change and continuity within Mayan families. *Ethos, 31*, 248–273.

Gaskins, S. (2008). Children's daily lives among the Yucatec Maya. In R. A. LeVine & R. S. New (Eds.), *Anthropology and child development: A cross-cultural reader* (pp. 280–288). Malden, MA: Blackwell.

Geertz, C. (1973). *The interpretation of cultures: Selected essays*. New York: Basic Books.

Gibbons, J. L. (2003, February). *The social and material worlds of contemporary Quiche Maya adolescents - a preliminary report*. Presidential Address presented at the annual meeting of the Society for Cross-Cultural Research, Charleston, SC.

Gibbons, J. L., & Stiles, D. A. (2004). *The thoughts of youth: Adolescents' ideal man and ideal woman in international perspective*. Greenwich, CT: Information Age Publishing.

Grönlund, H. (2013). Cultural values and volunteering: A cross-cultural perspective. In D. A. Vakoch (Ed.), *Altruism in cross-cultural perspective*. New York: Springer.

Huffmeijer, R., Alink, L. R. A., Tops, M., Bakermans-Kranenburg, M. J., & van IJzendoorn, M. H. (2012). Asymmetric frontal brain activity and parental rejection predict altruistic behavior: Moderation of oxytocin effects. *Cognitive, Affective, & Behavioral Neuroscience, 12*, 382–392.

Israel, S., Weisel, O., Ebstein, R. P., & Bornstein, G. (2012). Oxytocin, but not vasopressin, increases both parochial and universal altruism. *Psychoneuroendocrinology, 37*, 1341–1344.

Kraus, M. W., Piff, P. K., Mendoza-Denton, R., Rheinschmidt, M. L., & Keltner, D. (2012). Social class, solipsism, and contextualism: How the rich are different from the poor. *Psychological Review, 119*, 546–572.

Kruger, D. J. (2003). Evolution and altruism: Combining psychological mediators with naturally selected tendencies. *Evolution and Human Behavior, 24*, 118–125.

LaBouff, J. P., Rowatt, W. C., Johnson, M. K., Tsang, J.-A., & Willerton, G. M. (2012). Humble persons are more helpful than less humble persons: Evidence from three studies. *The Journal of Positive Psychology, 7*, 16–29.

Levine, R. V., Norenzayan, A., & Philbrick, K. (2001). Cross-cultural differences in helping strangers. *Journal of Cross-Cultural Psychology, 32*, 543–560.

Levine, R. V., Reysen, S., & Ganz, E. (2008). The kindness of strangers revisited: A comparison of 24 US cities. *Social Indicators Research, 85*, 461–481.

López, A., Najafi, B., Rogoff, B., & Mejía-Arauz, R. (2012). Collaboration and helping as cultural practices. In J. Valsiner (Ed.), *The Oxford handbook of culture and psychology* (pp. 869–884). New York: Oxford University Press.

Mathur, V. A., Harada, T., Lipke, T., & Chiao, J. Y. (2010). Neural basis of extraordinary empathy and altruistic motivation. *NeuroImage, 51*, 1468–1475.

Pawlik, K. (2012, July). *Ambulatory psychological assessment: Studying behavior in its natural ecology*. Paper presented at the 30th International Congress of Psychology, Cape Town, South Africa.

Pennebaker, J. W., Francis, M. E., & Booth, R. J. (2001). *Linguistic inquiry and word count LIWC*. Mahwah, NJ: Erlbaum.

Piff, P. K., Kraus, M. W., Côté, S., Cheng, B. H., & Keltner, D. (2010). Having less, giving more: The influence of social class on prosocial behavior. *Journal of Personality and Social Psychology, 99*, 771–784.

Rich, M. (2002). Show is tell. *Narrative Inquiry, 12*, 405–412.

Rogoff, B., Mistry, J., Goncü, A., & Mosier, C. (1993). Guided participation in cultural activity by toddlers and caregivers. *Monographs of the Society for Research in Child Development, 58*(8), 1–183.

Rosen, H. S., & Sims, S. T. (2011). Altruistic behavior and habit formation. *Nonprofit Management & Leadership, 21*, 235–253.

Rucker, D. D., Dubois, D., & Galinsky, A. D. (2011). Generous paupers and stingy princes: Power drives consumer spending on self versus others. *Journal of Consumer Research, 7*, 1015–1029.

Schlegel, A., & Barry, H., III. (1991). *Adolescence: An anthropological inquiry*. New York, NY: Free Press.
Schwartz, C. E., Keyl, P. M., Marcum, J. P., & Bode, R. (2009). Helping others shows differential benefits on health and well-being for male and female teens. *Journal of Happiness Studies, 10*, 431–448.
Selman, R. (1980). *The growth of interpersonal understanding: Developmental and clinical analyses*. New York: Academic Press.
Smith, S. W., Lapinski, M. K., Bresnahan, M. J., & Smith, S. L. (2013). Conceptual aspects of altruism in cross-cultural contexts. In D. A. Vakoch (Ed.), *Altruism in cross-cultural perspective*. New York: Springer.
Smith, S. W., Smith, S. L., Pieper, K. M., Yoo, J. H., Ferris, A. L., Downs, E., & Bowden, B. (2006). Altruism on American television: Examining the amount of, and context surrounding acts of helping and sharing. *Journal of Communication, 56*(4), 707–727.
Stellar, J. E., Manzo, V. M., Kraus, M. W., & Keltner, D. (2012). Class and compassion: Socioeconomic factors predict responses to suffering. *Emotion, 12*, 449–459.
Svetlova, M., Nichols, S. R., & Brownell, C. A. (2010). Toddlers' prosocial behavior: From instrumental to empathic to altruistic helping. *Child Development, 81*, 1814–1827.
Swain, J. E., Konrath, S., Brown, S. L., Finegood, E. D., Akce, L. B., Dayton, C. J., & Ho, S. S. (2012). Parenting and beyond: Common neurocircuits underlying parental and altruistic caregiving. *Parenting, Science and Practice, 12*, 115–123.
Visser, M. S., & Roelofs, M. R. (2011). Heterogeneous preferences for altruism: Gender and personality, social status, giving and taking. *Experimental Economics, 14*, 490–506.
Vollhardt, J. R., & Staub, E. (2011). Inclusive altruism born of suffering: The relationship between adversity and prosocial attitudes and behavior toward disadvantaged outgroups. *The American Journal of Orthopsychiatry, 81*, 307–315.
Wang, C., & Burris, M. A. (1997). Photovoice: Concept, methodology, and use for participatory needs assessment. *Health Education & Behavior, 24*, 369–387.
Wilson, N., Dasho, S., Martin, A. C., Wallerstin, N., Wang, C. C., & Minkler, M. (2007). Engaging young adolescents in social action through photovoice: The Youth Empowerment Strategies (YES!) project. *Journal of Early Adolescence, 27*, 241–261.

Chapter 5
Do We Really Like the Kind Girls and Animals?: Cross-Cultural Analysis of Altruism in Folktales

Alexandra Arkhipova and Artem Kozmin

5.1 Introduction

The kind of text we call the classical fairy tale already existed around 3,000 years ago. In the "Egyptian Tale of the Doomed Prince," dated around 1600–1300 BCE (Simpson, 1973, pp. 85–91), the main character, the prince, was doomed to be killed by a snake, a dog, or a crocodile. Although he could have killed his faithful dog, the prince refused to do so. Later on, he might have helped[1] another potential killer, the crocodile, and win a fight with a water spirit that was threatening him. According to those fragments, the prince performed one or more altruistic deeds and, thus, became a donor of altruistic acts. This is the oldest example of altruism in fairy tales we know of.

It is well known that humans and even animals (Boesch, Bolé, Eckhardt, & Boesch, 2010) can show an inclination to perform altruistic acts directed to other, even non-related, individuals. Social psychologists are interested in situations in which humans reveal altruism, whereas anthropologists are focused on the indirect benefits that a society can get from the altruistic behavior of its members. Most investigations of altruism are based on experimental observations that seek to clarify why people behave altruistically or egoistically and which environmental or cultural factors affect the degree and frequency of altruistic acts (see the detailed review by Piliavin & Charng, 1990).

The authors of this chapter, who are specialists in mythology and folklore, have a different approach to this question. We do not research real social behavior,

[1] The crocodile asked the prince for help, but unfortunately this papyrus does not state whether the prince saved his life by helping the crocodile.

A. Arkhipova (✉) • A. Kozmin
Centre of Typological and Semiotic Folklore Studies, Russian State University for the Humanities, Miusskaya sq. 6, Moscow 125 267, Russia
e-mail: alexandra.arkhipova@gmail.com; kozmin@gmail.com

D.A. Vakoch (ed.), *Altruism in Cross-Cultural Perspective*,
International and Cultural Psychology, DOI 10.1007/978-1-4614-6952-0_5,

altruistic or not. Instead, we focus on the imagination, that is, how people represent altruistic actions and transmit knowledge about them in the same types of text. More specifically, we try to understand altruistic behavior reflected in folktales.

Why should we study especially the *folktale* and not other folklore texts? We investigate the distribution of altruistic acts in folktales because they represent a widespread and easily transmitted genre of folklore; folktales have traditionally been retold orally, with a number of verbal stereotypes and clichés. Folktales are a rather ancient genre of folklore which, in modern European cultures, began to be recorded around 1700–1800 CE, with the German tradition having the largest number of plots. In ancient cultures—Egypt, India, Persia, and China—they began much earlier and were presented in a large number of ancient and medieval literary sources. Finally, the only way to perform a statistically comparative analysis of folklore texts is by using folktale material. The great value of folktales for comparative research is highlighted by the availability of a huge database, the The Aarne–Thompson Tale Type Index, which we describe below.

It is also important to note that the relationship between folktales and social reality on one side, and actual mythology on the other, is quite complicated. Folktales never reflect strictly ethnographic facts. For example, the most popular character in Russian fairy tales is Baba Yaga, a type of forest witch who is very ugly and sometimes kind and sometimes angry. If you do field work in any village in the Russian North, you easily find out that local folks do believe in *domovoj* (a spirit of the house) and *leshy* (the spirit of the forest); they worship and feed them, for example. But you will never find any actual mythological beliefs or magical practices connected to Baba Yaga. "It is only for scaring the kids," local people say if you ask them. And despite intense academic efforts to find ethnographic or archeological evidence of faith in Baba Yaga, no traces have been discovered.

A thorough review of the published literature on folklore studies reveals that a broad-based analysis of altruism in folktales has so far not been reported. We found only one interesting research paper about the relationship between tales and altruism, but its main approach and aim was different. That paper investigated the altruistic tales that parents tell their children, or in other words, which models of altruism parents prefer to teach their children. Moreover, that paper dealt with how parents choose which tales to tell and which social behavior to teach, rather than examining the folktales themselves (Palmer et al., 2006).

Another issue that remains to be evaluated is the correlation of altruism and folktales with social reality. We cannot automatically conclude that finding altruistic elements in folktales means that such altruistic events really existed in the historical past. We can infer only that people *have imagined them*. The prevalence of altruistic acts or motivations in the folktale deserves further investigation, which has special emphasis on the areal distribution of altruistic acts in folktale types.

In this chapter, we examine the different types of altruistic deeds discussed in folktales. The number of different types of folktales with altruistic content is lower than might be expected. As a rule, altruistic acts in fairy tales are directed to supernatural beings, namely, spirits and other magical creatures, not to humans. Finally, our research reveals that within many ethnic groups included in The Aarne–Thompson

Tale Type Index, the number of recorded altruistic tale types among Germanic cultures is slightly lower than in other areas, which can be explained by the influence of the Protestant ethic in this macroregion.

5.2 The Material of This Research: The Aarne–Thompson Tale Type Index and the Database of Folktale Types

The kind of text we call the *classical folktale* is very familiar to Western readers, especially through the Brothers Grimm's *Children's and Household Tales*. The Grimms used the word *Maerchen* as a genre label for their texts. The famous folklorist Stith Thompson in his *The Folktale* described Märchen as "a tale of some length involving a succession of motifs or episodes. It moves in an unreal world without definite locality or definite creatures and is filled with the marvellous. In this never-never land, humble heroes kill adversaries, succeed to kingdoms and marry princesses" (Thompson, 1977, p. 8). To be clear, however, this definition does not cover all Maerchen in the Grimms' book. For example, in *The Wolf and the Seven Young Kids*, there is no hero who marries a princess.

In 1910, Finnish scholar Antti Aarne published his famous book *Verzeichnis der Märchentypen*. Aarne and his followers (the historic–geographic School or the so-called Finnish School) had both a practical task (to collect the plots (*tale types* in Aarne's terminology) of European folktales) and a theoretical goal (roughly speaking, to trace the geographic origin of folktales). Although they never reached that goal, theirs was the first successful attempt to create a classification—or more precisely a catalogue—of folktales, including their areal distribution. Aarne's *Index*, revised by Stith Thompson in 1961 (Aarne & Thompson, 1961), offers a systematic method for describing folk narratives, tracing their origins, and suggesting paths of oral and literary transmission. This system has been severely criticized several times; see critiques and discussion by Propp (1958) and Dundes (1997).

However, Aarne–Thompson's Index version has a major problem with the geographical representation of ethnic groups: it is restricted to a limited areal distribution, namely, to European and Asian folktales, that is, the texts that had already been collected and described by folklorists at the time it was created. According to Stith Thompson, "strictly, then, this work might be called *The Types of the Folk-Tales of Europe, West Asia, and the Lands Settled by these Peoples*" (Aarne, Thompson, 1961, p. 7).

Hans-Joerg Uther, who updated the Aarne–Thompson system in 2004, noted that "it is self-evident that the revision must cover previously proposed enlargements of the tale-type system, and it must find a way to include the regions of Europe that have until now been underrepresented. Thompson barely considered Austrian and Swiss tales, neglected tales from Southern and Eastern Europe, and particularly Slavic tradition" (Uther, 2004). Nowadays, the Aarne–Thompson–Uther system (hereafter ATU) covers, in addition to "the types of the folk-tales of Europe, West Asia, and the lands settled by these peoples," the Eastern and South European,

Chinese and Siberian, and North African macroregions. We can see in the ATU, for example, Mayan tales, although in most cases they were borrowed from Spaniards. Also, some versions of tale types from regions not connected to Europe are today included in the ATU. For example, the system has now integrated the version of the tale type ATU 480 (*Kind and Unkind Girls tale*) found in the Kapingamarangi atoll (Federated States of Micronesia), a place that has never been settled by Europeans.

Folklorists have managed to obtain a very large sample of folktales. We lack an exact definition of the folktale genre, but that is a typical situation in many disciplines; for example, there is no exact definition of a novel in literary theory. But the ATU system provides a way to deal with the material: we label a text as a folktale if its plot can be described in terms of the ATU system.

Today's ATU Index allows users to design a special database that matches the plots, the areal distribution, first records of tales, and so on in a convenient way. Our own database, which provides the foundation for the results discussed in this chapter, now contains 141 ethnic traditions and 2,200 tale types but is not a strict copy of the printed version of ATU because ATU uses a lot of secondary literature. For example, the label "BP" in the plot descriptions of the printed version refers to the huge Bolte–Polivka volume of comments on the Brothers Grimm tales. This volume contains a great deal of information about versions of a given tale in different European traditions, although in the database these references are currently omitted.

5.3 Reciprocal Altruistic Acts in the Folktale

To clarify the process of investigating altruism in folktales, we first need to define altruism. There is some controversy about the right meaning of altruism and whether or not pure altruism exists (for a review of different theories, see Piliavin & Charng, 1990). Some scholars highlight reciprocal altruism (e.g., Palmer et al., 2006), while others argue that whenever the potential altruist already knows that his or her act can be reciprocated, no altruism is present at all. For example, if you meet an old lady begging for money in the street, you would probably give her a few coins. One could say that this is a truly altruistic act. Alternatively, one could also argue that if you do not help her, you will be tortured by a feeling of guilt because, for example, the lady reminds you of your grandmother. You benefit from this act because by giving her money you get a feeling of self-satisfaction, whereas not responding to her begging leaves you conscience-stricken. In this situation, no altruistic act has occurred.

Piliavin and Charng (1990, p. 29) state that "they have been unable to agree on a single definition of altruism." According to them, one of the main problems with finding the right definition involves "the relative emphasis on two factors: intentions and the amount of benefit or cost to the actor" (Piliavin & Charng, 1990, p. 29).

In this chapter, the concept of *reciprocal altruism*, with the above-mentioned restrictions, is used as our definition. One common tale type that features a plot incorporating reciprocal altruism is known as The Tale of the Kind and the Unkind Girls (ATU 480; Uther, 2004, I, pp. 281–283) as a typical example of plot with altruism. As Warren E. Roberts (1958, ix) succinctly summarized the core of the plot, "[t]he tale typically concerns two girls, one of whom is kind to a donor figure who rewards her kindness while the second is unkind to the same donor figure who appropriately punishes her." To provide a sense of the variations the tale can take, let us begin with the description of the kind girl, who has to work very hard and who is mistreated by her stepmother. This tale exists chiefly in two different forms:

1. A girl loses an object, such as a spindle, that is carried off by a river (or blown away by the wind), and she runs after it. The girl comes upon an old woman (witch) who asks her to do household tasks and to comb her hair. When the old woman lets the kind girl choose a box as her reward, she chooses a plain, small box rather than an attractive, large one.
2. A girl falls into a well (or is pushed in, or jumps in after an object she has dropped), or she follows a rolling cake (or ball of yarn, etc.). She encounters various animals, objects, or people who ask her for help. For example, a cow wants to be milked, an old man or woman to be fed or to have hair lice removed from their head, an oven wants to be emptied of bread, or an apple tree wishes to be shaken. The girl helps them all. Next, she arrives at the house of an old woman (or supernatural beings such as giants) where she is assigned tasks. For instance, she has to do household or farm work, she needs to feed or louse a demon, she must comb the hair of fairies, or she has to carry water in a sieve or wash black wool until it is white. The kind girl is very helpful and hardworking. In the end she is rewarded with gold (or jewels, etc.), she becomes more beautiful, or she is allowed to choose her reward and does so very modestly, but the reward changes into wealth when she reaches home.

In some variants, she is pursued by the being she met earlier and is helped again by animals, plants, or other objects. In some versions, she is forbidden to enter a certain room. When she breaks this rule, she becomes covered with gold and flees the room. The grateful animals, objects, and people of her outward journey help her escape from her pursuer.

After she arrives home, her envious stepmother sends her own daughter on the same journey. This girl goes through the same situations but she refuses help, disobeys, and is unkind. She is punished severely (e.g., frogs fall from her mouth; horns grow on her head; or she is disfigured, beaten, or killed). In some variants, the stepmother is also punished. In the end, the kind young woman often marries a prince.

Let us consider this Kind and Unkind Girls tale type as a typical example of a plot with altruism. We can notice special features of altruistic acts commonly found in folktales. First, altruistic acts in folktales take the form of reciprocal altruism. As stated by Palmer et al. (2006, p. 240), "Reciprocal altruism is an evolutionary concept used to explain altruism among individuals who may or may not be closely related.... [A]ltruistic acts could be favored by natural selection when they are likely to be reciprocated in a form that is at least as valuable to... the altruist as the original act was costly." For example, just because of her good nature, the kind girl helps an apple tree, and later she is assisted by the same tree. Moreover, in this reciprocal altruistic act, we see the presence of a donor and a recipient: the kind girl is the donor and apple TREE is the recipient.

5.4 How Many Different Altruistic Tale Types Exist?

As we have already mentioned, our analysis is based on a special sample from our database of tale types, which includes 141 ethnic traditions and 2,200 tale types (plots) (see Table 5.1). The sample of plots containing any altruistic acts was based on experts' evaluation. Of course, several cases are still questionable, because we could not evaluate if this deed was altruistic or not. Although other scholars might obtain slightly different results, we are convinced that our results reflect reality.

Before going further, we need to ask, how many altruistic tale types exist? Are altruistic tale types rare or common? We believe that most readers would predict a large number of different altruistic plots. However, we found a very low number—only 72 plots among 2,200 tale types—containing some altruistic ideas and motifs.

Why may most people have such expectations? This might be the case because several plots with altruistic topics are widespread. For example, the Kind and Unkind Girls tale type can be found in more than 900 versions (Roberts, 1958) (Fig. 5.1).

The second popular tale type among the altruistic group, which is depicted in Fig. 5.2, is *The Magic Ring* (ATU 560): A boy buys a dog, a cat, and a snake to save them from death (he rescues a snake from burning, for example). In return, he receives a magic ring (or stone) from the king of the snakes (who is the father of the rescued snake), which will grant all wishes. With this ring, the boy builds a magic castle and performs impossible tasks (e.g., building a bridge of glass or a church of wax). Then he marries the princess, but she has a lover, and so the adventures of our hero continue.

In the ten most popular fairy tales of any kind, there are at least two altruistic plots (Figs. 5.1 and 5.2). The ten most popular tale types that are specifically altruistic are listed in Table 5.2. These tale types are found in a range of different traditions. We believe this explains the impression that altruistic tales occur with higher prevalence than they really do; only 72 of 2,200 plots are altruistic, but these relatively few altruistic tales are widespread, and almost everybody knows them in many cultures.

Table 5.1 Number of total plots and altruistic plots in ATU

Total number of ethnic traditions in ATU	141
Total number of plots	2,200
Number of altruistic plots	72

Fig. 5.1 Areal distribution of the Kind and Unkind Girls tale type (ATU 480)

Fig. 5.2 Areal distribution of the Magic Ring tale type (ATU 560)

Table 5.2 The ten most prevalent altruistic tale types

Title of the altruistic plot	Number of traditions including this altruistic tale type
The Magic Ring (ATU 560)	82
Kind and Unkind Girl (ATU 480)	82
The Grateful Animals (ATU 554)	81
Bird, Horse, and Princess (ATU 550)	79
The Clever Horse (ATU 531)	76
The Faith of the Lion (ATU 155)	74
The Man Who Understands Animal Languages (ATU 670)	72
One-Eye, Two-Eyes, Three-Eyes (ATU 511)	70
The Smith and the Devil (ATU 330)	68
The Faithless Sister (ATU 68)	68

5.5 Donors and Recipients of the Altruistic Acts in Fairy Folktales

Traditional cultures are built on a system of reciprocal gift exchange (Mauss, 1966). Such a system makes the existence of reciprocal altruism both possible and highly valuable, not only directly, in a physical way, but also indirectly, in a symbolic way. The altruistic behavior of individuals is rather typical within a traditionally oriented society. Different kinds of reciprocally altruistic acts can be found either directed from a person to his/her relatives or among non-related individuals.

It is logical to expect the same situation in folklore, where we might anticipate the abundance of altruistic acts and the existence of gift exchange. However, in folklore the situation with altruism is completely different. In the fairy world, two human beings almost never become the donor and the recipient of an altruistic act as it happens in normal life. For example, in any human society an adult would always adopt an orphan baby. In contrast, in the fairy world, children who have lost the support of adults are often simply abandoned by humans and then are luckily found by a supernatural creature that takes care of them.

Plots in which one person helps another are astonishingly rare: only in two cases within our database (ATU 516C,[2] 864[3]) does a human character help another person.

[2] *Amicus and Amelius.* Two friends who look exactly alike assist each other in time of need. When one of them, who in older versions is named Amelius, is challenged to a sword fight, the other, Amicus, who is a better swordsman, takes his place. Amelius remains behind with Amicus's wife and puts a sword in the middle of the marriage bed at night. Amicus wins the fight for his friend. Later, Amicus contracts leprosy. An angel tells them that he will be healed if he bathes in the blood of Amelius' children. Amelius cuts off their heads to save his friend. The children are restored to life (ATU, 516C).

[3] *The Falcon of Sir Federigo.* The nobleman Federigo Alberighi loves Monna Giovanna, but she does not return his love. He spends all of his wealth to woo her, until he is left with only his falcon, which is one of the best of its kind. When Giovanna's sick son wants the falcon, she goes to ask for it. In order to gain her favor, Federigo unwittingly serves her his precious falcon to eat. When she makes her request known it is too late, but she is so impressed by his generous attitude that she changes her mind and chooses him as her husband (ATU, 864).

And, except in those two examples, it is rather difficult to find a situation where a man helps his relatives. The opposite situation, where humans affect each other, is quite common. For instance, readers may remember a number of tale types about older brothers who try to abandon or to kill a younger sibling, as in the case of Joseph and his brothers.

Who are the donor and the recipient of altruistic acts in folktales? Instead of *animal tales*, which are allegorical and feature at least two creatures, in *fairy folktales*, which highlight opposition between natural and supernatural worlds, the altruistic act occurs between the main character, who is a human, and an animal/magical creature or a person with some hidden or obvious supernatural power. The nonhumans (animals or other magical creatures) can be both recipients and donors of altruistic acts. When Ivan feeds the Grey Wolf with his horse, and the Wolf becomes his helper in return, the Grey Wolf is initially the recipient of Ivan's altruistic act but then becomes the donor. The same applies to the traveler who saves animals from their death and is rewarded by the animals in return.

In almost all cases in which magical creatures are donors at the beginning of a story, the first element of the donor–recipient relationship exists but is reduced. For example, Ivan meets an ugly old lady in the forest and, unlike his brothers, treats her politely. In return, the lady gives him wise advice or the right directions. In this example, the old lady was initially the recipient of an unclear altruistic act (Ivan's courtesy to her) and only then did she become the donor.

Sometimes, the protagonist must even be taught how to ask supernatural creatures for help. Tsarevich Ivan is advised by a helper (for instance, his late father) about what he should say to Baba Yaga, the forest witch, so that he can be rewarded instead of being eaten.

The worldview of the fairy tale contrasts dramatically with the rules of traditional society. In a fairy tale, an altruistic act whose donor is a human is almost never directed to another human but almost always addressed to a representative of the supernatural world, and vice versa. This tells us that the moral of fairy tales concerns how to contact spirits—feed them or behave in an especially polite way—in other words, *treat supernatural powers or magicians well and they can help you in return!*

5.6 When Altruistic Acts in the Folktale End Badly

We should remember that Aarne–Thompson–Uther's Tale-Type Index organizes plots into very broad categories like *animal tales*, *fairy tales*, and *religious tales*. Within each category, folktale types are subdivided by motif down to the level of individual tale types. A typical example of an animal tale is "Ice House," in which a fox builds himself a fine house of ice. A hare (or wolf, fox, bear, sheep, or goat) builds a house of wood (or stone, iron, grass, or wool). When summer comes, the fox's house melts, and he goes to live in the hare's house, and gradually the fox pushes the hare out.

Table 5.3 Winning and losing strategies in altruistic animal and fairy tales

	Animal tales	Fairy and religious tales
Number of tale types with winning strategy	3	56
Number of tale types with losing strategy	12	1
Total number of altruistic plots	15	57

Fig. 5.3 Areal distribution of the Ice House tale type (ATU 43)

From the main character's point of view, altruistic acts in animal and fairy folktales can be part of a winning or losing strategy. The story about the kind girl and the apple tree is an example of a winning strategy because, as a result of her altruistic act, she receives wise advice from the tree and wins, whereas her sister refuses to help the tree and therefore loses. A strategy of winning altruism is popular among fairy tales and the so-called religious tales (59 cases). But 13 tale types, most of which are animal tales and only a few of which are fairy or religious tales, include altruistic acts that end badly (Table 5.3). Such an example is the Ice House tale type (ATU 43; see Fig. 5.3).

One of the differences between fairy tales and animal tales is in the divergent impact of altruistic acts. In fairy tales, the altruistic act is a trigger that starts the protagonist's quest: at the beginning the actor helps magical animals, and afterwards he receives benefits. In animal tales, the altruistic act is an initial part of the trickster's work, which very often has a bad ending for the donor of the altruistic act.

5.7 Who Does Not Like Altruistic Folktales?

One can predict that the number of altruistic plots depends on the total number of plots in a given tradition. A first look at the data presented in Table 5.4 apparently confirms this impression. However, a deeper look into the areal distribution of altruistic plots reveals a slightly different picture. To analyze these data correctly, we need to see if there is a correlation between the areal distribution of altruistic plots and other important parameters. That is not a simple task, requiring a rather complicated statistical procedure.

The first step is to find the fraction of altruistic tale types within the general tradition. When we started to calculate this fraction, we confronted a problem typical for statistical analyses in social sciences: not all traditions have the same number of tale types. This problem is especially acute for traditions with a low number of tale types. For example, if in one of those traditions there is only one altruistic tale type out of a total of three entries, the probability of an error due to chance is high. To avoid this error, we validated our results by performing a principal component analysis (PCA), which is a mathematical procedure that allows us to convert a set of observations of possibly correlated variables into a set of values of linearly uncorrelated variables

Table 5.4 Top 20 ethnic groups having plots containing altruistic acts

Ethnic group	Total number of plots	Number of altruistic plots
German	1,408	48
Latvian	1,122	45
Finnish	1,140	45
Hungarian	1,029	40
Ukrainian	860	40
French	837	39
Greek	877	37
Russian	815	37
Polish	754	36
Irish	741	36
Spanish	1,042	36
Italian	847	35
Lithuanian	923	34
Jewish	882	34
Czech	571	33
Swedish	727	33
Belarusian	622	33
Gipsy	504	31
Bulgarian	854	31
Slovakian	520	30
Chinese	540	30
Catalonian	604	28
Estonian	778	28
Indian	717	27

called principal components. The transformation is defined in such a way that the first principal component has the largest possible variance (Shaw, 2003).

In our database, which includes 2,200 tale types and 141 ethnic groups, we identified a total of ten components, or factors, for assessing the traditions. Tale types that are likely to appear together in the same ethnic group belong to the same factor. For example, Factor 5 is labeled "Iberian" because it contains tale types that are found with a high probability in Spanish, Catalonian, French, and Portuguese traditions, and nowhere else. Tale types 159A, 285E, 186, 2B, and 157C* are found only in the Iberian group. One example of a Factor 5 tale is ATU 189—*The Monkey and the Nut.* This tale, in which a monkey throws a nut away because it has a bitter shell, and overlooks the edible kernel, is present only in the Spanish and Catalonian traditions. In such a way, in the European set of tale types, several components can be highlighted. This allows us to distinguish which types are specific to each region. Consequently, the principal component analysis allows us to distinguish tale types that have specific geographical distributions.

In a next step, we looked for correlations between the fraction of altruistic tale types and various factors. We found several significant (but weak) positive and negative correlations. When we see a negative correlation between the fractions of altruistic tale types and a factor, it means that the more geographically specific tale types exist, the lower the chance these will include altruistic tale types. Conversely, a positive correlation means that the more geographically specific tale types exist, the higher the probability of finding altruistic ones among them.

After conducting our analyses, we found a negative correlation between the "North Germanic–East European" factor and the altruistic plots fraction: it is very significant ($p<0.05$) but weak ($r=-0.184$). In this factor German, Frisian, Dutch, American, English, Flemish, and Danish traditions have maximum factor scores. On the other hand, Bulgarian, Estonian, Latvian, Belarusian, Lithuanian, Russian, and Ukrainian have minimum factor scores. Therefore, Factor 2 allows us to distinguish between North Germanic and East European traditions. It should be noted that this distinction is based on about 400 plots. In simple terms, the North Germanic set of traditions dislikes altruistic tale types, whereas the East European one, in contrast, likes to use them. (In more technical terms, this principal component explains 19 % of the variance.) However, this relationship does not depend on the absolute number of tale types in given traditions; we talk about only geographically specific tale types that distinguish these two major areas!

The second correlation is also negative ($r=-0.14$), being weak but significant ($p<0.05$). It is associated with Factor 3, differentiating "South" traditions (Spanish, Jewish,[4] Indian, Egyptian, Greek, Portuguese, Italian), which have maximum factor scores, from "Baltic" traditions (Lithuanian, Swedish, Estonian, Latvian, Finnish), which have minimum scores. It means, again, that "Baltic" set of tale types is "less altruistic" than the "South" ones.

In both cases we deal with Germanic and Baltic cultures (German, Frisian, Dutch, American, English, Flemish, Danish, Swedish, Estonian, Latvian, and Finnish),

[4] Near East Jewish communities

where the parents preferred to tell their children stories in which the main character saves himself by his own efforts rather than sitting and waiting for altruistic help from magical creatures. That is a possible reason why the collectors of classic folklore there recorded a slightly lower number of altruistic tale types than in other regions during the last 200 years, even if the total number of recorded folktales is high. It is noteworthy that these are countries (with the partial exception of Flanders and South Germany) where Protestantism was widely spread during the last 300 years.

5.8 Conclusion

Reciprocal gift exchange provides a foundation for the reciprocity we see in altruistic acts of individuals from traditional societies, whether the altruism is directed toward kin or unrelated individuals. In folklore, the total number of different altruistic plots throughout the world is not especially high, although the types that do exist are sufficiently widespread that they are well known by people from geographically diverse locations. Altruistic acts in fairy tales serve to initiate the protagonist's mission, while they more often harm the giver in animal tales, which are allegories of humans engaging with one another. That is, altruism in fairy tales is likely to be profitable to the donor. In contrast, altruism in animal tales is more often than not unprofitable for the altruist. Unlike interactions between real human beings, in fairy tales two humans are almost never the giver and receiver of altruistic acts. Instead, supernatural creatures care for humans in fairy tales, and they can also be recipients of altruistic acts.

The geographical distribution of altruistic tale types may correspond to cultural differences between Catholic and Protestant regions. Protestant beliefs are based on the individual's own abilities, and they support the rational pursuit of economic gain (Weber, 1930). We hypothesize that Protestantism does not admire people who passively wait for help from others—even in folktales.

Acknowledgment We are truly grateful to Prof. Andrei Korotaev and Dr. Carlos Martín Medina for their help.

References

Aarne, A., & Thompson, S. (1961). *The types of the folktale: A classification and bibliography*. Helsinki: The Finnish Academy of Science and Letters.

Boesch, C., Bolé, C., Eckhardt, N., & Boesch, H. (2010). Altruism in forest chimpanzees: The case of adoption. *PLOS ONE, 5*(1), e8901.

Choi, J. K., & Bowles, S. (2007). The coevolution of parochial altruism and war. *Science, 318*, 636–640.

Dundes, A. (1997). The motif-index and the tale type index: A critique. *Journal of Folklore Research, 34*(3), 195–202.

Mauss, M. (1966). *The gift: Forms and functions of exchange in archaic societies*. London: Cohen and West.
Palmer, C. T., Wright, J., Wright, S. A., Cassidy, C., VanPool, T. L., & Coe, K. (2006). The many manipulations of Morty Mouse: Children's stories and the parental encouragement of altruism. *Journal of Anthropological Research, 62*(2), 235–257.
Piliavin, J. A., & Charng, H.-W. (1990). Altruism: A review of recent theory and research. *Annual Review of Sociology, 16*, 27–65.
Propp, V. (1958). *Morphology of the folktale*. (L. Scott, Trans.) Bloomington, IN: Research Center, Indiana University.
Roberts, W. E. (1958). *The tale of the kind and the unkind girls: AA-Th 480 and related titles*. Berlin: W. de Gruyter.
Shaw, P. (2003). *Multivariate statistics for the environmental sciences*. London: Arnold.
Simpson, W. K. (Ed.). (1973). *The literature of ancient Egypt: An anthology of stories, instructions, and poetry* (R. O. Faulkner, E. F. Wente, Jr., & W. K. Simpson, Trans.). New Haven, CT: Yale University Press.
Thompson, S. (1977). *The folktale*. Berkeley, CA: University of California Press.
Uther, H.-J. (2004). *The types of international folktales. A classification and bibliography*. Parts I-III. Helsinki, Finland: The Finnish Academy of Science and Letter.
Weber, M. (1930). *The Protestant ethic and the spirit of capitalism*. London: Unwin Hyman.

Chapter 6
Cultural Values and Volunteering: A Cross-Cultural Perspective

Henrietta Grönlund

6.1 Introduction

No single, simple, and objective definition for volunteering exists. The lack of specificity and the boundaries for this phenomenon have been discussed from several viewpoints, including different styles of voluntary action, different motives for such actions, the beneficiaries, and the organizers of voluntary action (Cnaan, Handy, & Wadsworth, 1996; Rochester, Paine, Howlett, & Zimmeck, 2010). The act of volunteering is known throughout the world, but the word is not (2011 State of the World's Volunteerism Report, 2011). Cultural contexts influence what is defined as volunteering, which makes cross-cultural approaches challenging. The status of volunteerism in different countries and different linguistic concepts influences this (Dekker, 2002). Nevertheless, Cnaan, Handy, and their colleagues (1996; Handy et al., 2000) have shown that in spite of inconsistencies in defining volunteering in different countries, there is a reasonable consensus among respondents on what constitutes volunteering. In this chapter, we adopt a relatively simple definition based on features which are recognized across cultures: Volunteering is giving one's time freely and without financial reward to help other people or a cause in an organized manner. Helping one's family members or friends is not volunteering. Our definition includes volunteering in a variety of activities and fields ranging from board memberships to protecting endangered animals, from rebuilding communities after disasters to taking action for social justice, and from sports to welfare.

Volunteering has been the object of numerous studies (for a summary, see Musick & Wilson, 2008). The impact of sociodemographic backgrounds, religion, and social factors have been scrutinized in different contexts, along with the motives and psychological functions of volunteering (e.g., Clary et al., 1998; Cnaan & Goldberg-Glenn, 1991; Musick & Wilson, 2008; Yeung, 2004). Volunteering is often regarded

H. Grönlund (✉)
University of Helsinki, Helsinki, Finland
e-mail: Henrietta.gronlund@helsinki.fi

D.A. Vakoch (ed.), *Altruism in Cross-Cultural Perspective*,
International and Cultural Psychology, DOI 10.1007/978-1-4614-6952-0_6,

as altruistic behavior but has been often studied from an egocentric approach where individuals are seen to act foremost to fulfill their own needs and interests. This approach has been common in many disciplines that study volunteering (Haski-Leventhal, 2009). Both self-directed and altruistic motives have been seen to motivate volunteering. These motives do not exclude each other but are usually combined, and motivation can have different emphases at different stages of life or the volunteering process. Research on volunteerism shows that common self-serving motives are related to one's career or learning, social motives related to social connections or rewards from others, and enhancement motives, where volunteering provides self-esteem and feelings of being needed. Other-oriented motives for volunteering include different explications of altruistic motivation, empathy, justice, and a will to help (Clary et al., 1998; Musick & Wilson, 2008).

Both self and other-oriented motives are connected to personal values. Values are defined as relatively stable guiding principles in the life of an individual. They include both emotional and rational elements and are often components of the motivational structures of individuals (Rokeach, 1973; Schwartz, 2007, p. 712). Thus, values guide most actions of individuals, including volunteering. The influence of values can be acknowledged or unacknowledged. The *2011 State of the World's Volunteerism Report* by the UN (2011) maintains that solidarity, compassion, empathy, and respect for others are ingrained in many communities around the world and are often expressed through the giving of one's time. Volunteerism can also be an expression of the desire to act on one's feelings about justice and fairness and to foster social harmony (2011 State of the World's Volunteerism Report, 2011). Values are a common field of study in relation to volunteering. Altruistic, humanitarian, and religious values have been strongly linked with volunteering in empirical studies, although we can also find connections with self-enhancing values. A self-enhancing value orientation can be seen as associated with self-serving volunteering motives and a value orientation directed toward self-transcendence is logically associated with volunteering motives related to the well-being of others (Bekkers, 2005; Clary et al., 1998; Cnaan et al., 2012; Grönlund, 2011; Musick & Wilson, 2008). Nevertheless, values are always dynamic and multidimensional, and straightforward causalities should be avoided (Schwartz, 1992).

Societal contexts provide the sociopolitical models that regulate the requirements and possibilities for volunteering (Stadelman-Steffen & Freitag, 2011). Furthermore, the substantial differences in values and norms between cultures lead to different emphases in acknowledged motives for volunteering (Hofstede, 2001; Inglehart, 1997). Cultural values are defined as the implicitly or explicitly shared abstract ideas about what is good, right, and desirable in a society. In every culture, social institutions such as schools and the justice system reproduce these ideals and teach individuals what is appropriate for different situations based on these norms. Cultures are never completely uniform but include subcultures and groups that have different values. When power relations between the majority and subcultures shift, the cultural orientation, cultural values can also change. Nevertheless, such change happens gradually. Cultural values are measured through the mean importance of different values among individuals in a particular country. Although each individual has his or her own value priorities, and there can be significant differences among individuals within the same cultural context, this does not affect the mean of the importance of different values. Thus, the average value priorities of individuals in a particular country express the dominant cultural value directions (Hofstede, 2001; Schwartz, 1999, 2007). Some cultural values are more self-serving and direct people toward self-enhancing action, whereas other cultural values are more altruistic and direct them toward self-transcending action (Schwartz, 2007). Similarly, individualistic and communal cultures direct individuals to different action or different motives for a similar action (Hofstede, 2001). As Smith, Lapinski, Bresnahan and Smith (2013) note in Chap. 2 of this volume, behaviors that would seem altruistic in an individualistic culture would not necessarily be seen as such in a communal culture. According to them, different composites of altruism are likely to be coded differentially across communal and individualistic cultures. Also, Gibbons discusses the importance of considering the social context in understanding individuals' views of altruism in the fourth chapter of this book (Gibbons, 2013). The current chapter discusses influence of cultural contexts, especially cultural values in volunteering and its motives. It will also look at theoretical viewpoints to cultural values and empirical studies that focus on volunteering and values or volunteering in cross-cultural perspectives.

6.2 Cross-Cultural Viewpoints to Volunteering

Most extant research on volunteerism examines motivation, its connection with values and other volunteering-related issues in a single nation partly because definitional and contextual variations make cross-cultural comparisons problematic (Anheier & Salamon, 1999; Dekker & Halman, 2003; Musick & Wilson, 2008). Despite cultural differences, similar motives for volunteering as well as similar roles of values in volunteering have been discovered in studies conducted in various individual countries and cultural settings (Musick & Wilson; 2011 State of the World's Volunteerism Report, 2011). Researchers have also sought to overcome national boundaries in volunteering research with cross-national empirical studies. Such studies have highlighted a number of factors that may influence the level of volunteerism. These factors can be divided into structural factors and cultural factors. Structural factors include a nation's political system and experience with democracy, the degree to which it is a welfare state, its level of economic development and income inequality, and the age and ethnic structure of the society. Cultural values that influence volunteering include, for example, the level of individualism, religiosity, and trust within a country. Structural and cultural factors influence each other and are in continuous interplay (Hodgkinson, 2003; Musick & Wilson, 2008; Ruiter & De Graaf, 2006; Salamon & Sokolowski, 2003).

The influence of a nation's political regime on volunteering has interested researchers (Salamon & Anheier, 1998). Some have argued that the voluntary sector is stronger in countries with a longer experience of democracy (Curtis, Baer, &

Grabb, 2001; Halman, 2003). These countries have a long history of formal volunteerism, documented back to the nineteenth century. Although both liberal and social democratic regimes experience higher rates of volunteerism (Curtis et al., 2001), liberal regimes exhibit the highest rates (Parboteeah, Cullen, & Lim, 2004). Other people argue that volunteerism flourishes in countries that have been deprived of democracy. In Korea, for example, volunteerism and civil society have grown significantly since the 1990s with strong governmental support (Ziemeck, 2003). Countries with conservative regimes, such as Japan, have levels of volunteerism similar to transitional democracies but for different reasons. Governments of conservative countries do not encourage volunteerism, and transitional democracies have poorly developed civil societies. Both factors serve to limit the rate of volunteerism (Musick & Wilson, 2008). Anheier and Salamon (1999) have posited that the size of a country's nonprofit sector is the most significant structural factor affecting voluntary participation.

Communist countries and former communist countries experience (or have experienced) volunteering differently. For example, in Soviet regimes, volunteerism was state organized and compulsory (Musick & Wilson, 2008), which raises questions about whether it is volunteering or simply enforced unpaid work. In Croatia, formal volunteering was state organized but was not compulsory, and many individuals volunteered informally among family and friends. While this means that former communist countries in Eastern Europe have lower national rates of formal volunteering when compared to older democracies, the level of volunteerism is increasing, in contrast to Western European countries, where the rates are either static or in decline (Musick & Wilson).

The political system is also related to the welfare model of each country. Welfare models differ according to liberal, conservative, and socialist political principles, and different welfare ideologies are associated with different value positions. The way in which responsibility for welfare is allocated among state, market, and families can influence the needs, norms, and motives for volunteering in different countries (Billis, 1993). Researchers have argued, however, that a political system has a more significant impact on the type of volunteering individuals engage in within that country, rather than on participation rates. For example, in social democratic countries, volunteering is mostly in expressive activities (i.e., recreation), whereas in liberal and corporatist states, volunteering is more service-oriented (Salamon & Sokolowski, 2003).

Religion is a central factor in shaping cultures and nations. Indeed, organized religion affects both structural factors and cultural values within a country. In addition, religion takes on different meanings, values, and thus motives in different cultures. Most religions encourage helping other people and often offer possibilities for this, for instance, through their churches, missions, and kibbutzim. Thus, religion is often connected to both value-based and socially motivated volunteering. Nevertheless, there are differences in the ways in which religions operate in different cultures and their connection to volunteering. For instance, synagogues in Israel do not encourage volunteering but are strictly places for prayer. By contrast, in most Christian cultures, the churches provide many opportunities to volunteer and are viewed as strong builders of social capital (e.g., Yeung, 2004). Eastern religions often encourage altruistic behavior extended to all living creatures, as discussed by Gupta (2013) in Chap. 8 of this volume. The viewpoint of religions concerning gender and gender roles, the sociopolitical context in which the religion operates, and the status and influence that religion has in a culture are also connected to the role religion has in promoting volunteering in different nations.

Previous research (e.g., Ruiter & De Graaf, 2006; Wuthnow, 1991; Yeung, 2004) has shown that religious activity such as attending services is associated with increased volunteering. Religious beliefs also influence the meaning of volunteering in the lives of individuals. Social networks and questions of identity draw people into volunteering because many religions promote altruistic and pro-social behavior and offer possibilities (and often norms and expectations) to volunteer. Curtis and colleagues (2001) noted in their cross-national study of 33 democratic countries that membership in voluntary associations was higher in multidenominational Christian or predominantly Protestant religious populations. Also, Ruiter and De Graaf showed that differences between religious and nonreligious individuals in volunteering activity are smaller in devout countries compared to countries with a secular culture. A devout national context therefore had a positive effect on volunteering rates among both religious and nonreligious individuals (Ruiter & De Graaf, 2006). This highlights the cultural influence of religion on volunteering.

Despite the differences in definitions and practices related to volunteering in different contexts, we also see similarities. Hodgkinson examined trends across 47 countries and found that in each country, the rate of church attendance, membership in voluntary associations, and socializing through networks and political engagement were all significantly higher for volunteers than for non-volunteers (Hodgkinson, 2003). Handy et al. (2000) examined whether in spite of these definitional inconsistencies, people had similar perceptions of what constitutes volunteerism across five different countries: Canada, Italy, India, the Netherlands, and the USA. The key finding from this study was that there was a reasonable consensus among respondents on what volunteering was, although there was less agreement over more marginal forms of volunteerism. Cross-national studies of volunteerism are, however, in their infancy (Musick & Wilson, 2008) and have yet to attribute differences in volunteer participation to specific factors.

6.3 Cultural Values and Volunteering

Individualist cultures are those where the ties between individuals and groups are loose, whereas in collectivist societies people are integrated into close-knit groups from birth (Hofstede, 2001). Countries with high levels of individualism, such as Canada, the USA, and the UK, also exhibit high levels of volunteer participation (e.g., Grönlund et al., 2011). While volunteerism may appear to have a link with collective action—for example, South Korea has a high rate of volunteering and very low individualism score—countries with higher individualism scores such as Canada, the USA, and the UK also have high levels of volunteerism. Inglehart (2003) argues that volunteerism is related to the level of industrialization within a country, and as a country moves from an agrarian economy with traditional values toward an industrial economy with secular-rational values, the rate of volunteering increases. As most surveys have sought to capture data on formal volunteering, or volunteering through an organization, rather than informal volunteering, or helping out within one's own community (Carson, 1999), it may be that more collectivist societies have higher incidences of informal volunteering. Informal volunteering has typically been higher than formal volunteering among non-White communities in Western countries, including Australia, the USA and the UK (Musick, Wilson, & Bynum, 2000).

Schwartz (2007) has developed seven ideal types or clusters of values on which cultures can be compared. These value types are located on three polar dimensions. The individual elements and the dimensions of these value types are presented below (Schwartz, pp 4–6):

1. *Autonomy and conservatism*
   Autonomy: Divided in two kinds of autonomy—(a) intellectual autonomy, desirability of individuals to pursue their own ideas and intellectual directions independently (curiosity, broadmindedness, creativity) and (b) affective autonomy, emphasis on individuals to pursue affectively positive experiences independently (pleasure, exciting life, varied life)
   Conservatism: Emphasis on maintaining the traditional order (social order, respect for tradition, family security, wisdom)
2. *Egalitarianism and hierarchy*
   Egalitarianism: Emphasis on voluntary commitment to promoting welfare of others (equality, social justice, freedom, responsibility, honesty)
   Hierarchy: Emphasis on the legitimacy of unequal distribution of power, roles, and resources (social power, authority, humility, wealth)
3. *Harmony and mastery*
   Harmony: Emphasis on fitting harmoniously with the environment (unity with nature, protecting the environment, world of beauty)
   Mastery: Emphasis on getting ahead through active self-assertion (ambition, success, daring, competence)

Egalitarianism is the cultural value that comes closest to altruistic emphases. Schwartz (2012) has detected statistically significant connections between egalitarian cultural values with different structural and cultural features, especially when combined with emphases on autonomy. For example, they are positively connected with gender equality and higher governmental investments in health and education. Egalitarian cultural values have also been connected positively with the level of democracy in a country (Schwartz) Thus, many structural factors connected with volunteering rates and orientations are also connected with cultural values. Cultural values can be seen as the foundation which influences volunteering both directly and through structural factors.

Grönlund and colleagues (2011) examined the motivation of volunteers in 13 different countries and the contexts of their cultural values according to Schwartz (2007).

Because a myriad of factors influence volunteering ranging from opportunities to regulations, and needs to norms, it is difficult to formulate hypotheses regarding the influence on cultural values in volunteering rates. The problems with such an approach have also been detected in previous research, as summarized above. Instead, as discussed above, volunteering motivation is strongly connected with values. This makes volunteering motivation an interesting area of study in relation to cultural values. Motivation is also a central component in defining altruistic behavior (see Chap. 2 of this volume by Smith et al. (2013)). The above-discussed contextual factors of the 13 countries studied by Grönlund et al. (2011) are summarized in Table 6.1, noting similarities and differences. The table orders the countries according to the adult participation rate in volunteerism; the results from the survey of student volunteering appear in Column 3. Comparable data across all 13 countries are included where available. The central cultural values according to Schwartz (2007) for each country in the sample are presented in Table 6.1. Also, Hofstede's individualist-collectivist scale is included as it has been accepted as a valid measure in cross-cultural studies (Schimmack, Oishi, & Diener, 2005).

The survey collected data from over 600 students in 13 countries using a unified survey instrument. More than 9,000 students were surveyed from Belgium, Canada, China, Croatia, England, Finland, India, Israel, Japan, Korea, the Netherlands, New Zealand, and the USA (for details on data collection, method of analysis, and results, see Grönlund et al., 2011). Table 6.2 presents a descriptive analysis of five dimensions of volunteering motivation.

Students expressed the strongest support for altruistic and learning motives for volunteering across all countries. Altruistic motives in this study included the following reasons for volunteering: *because it is important to help others* and *to work for a cause that is important*. Learning motives included the following: *to learn more about the cause for which one is volunteering, volunteering gives one a new perspective, volunteering makes one feel better,* and *to make new friends*.

Career and social motivations came third and fourth and protective reasons for volunteering were considered least important. Career motives included the following: *to put it on their C.V. (resume) for admission to a higher level educational institute, to put it on their C.V. (resume) when applying for a job, volunteering helps get a foot in the door at a place where one wants to get paid employment,* and *to make new contacts that might help a business career*. Social motives included the following: *friends volunteer, were advised to do so by career counselor or family,* and *people close to one influence them to volunteer*. And finally protective motives included the following: *Doing volunteer work relieves some of the guilt over being more fortunate than others,* and *volunteering is a good escape from one's own troubles*.

Thus, altruism is a central motive that individuals give for their volunteering despite their cultural context. Notwithstanding the similarity in these general rankings, important country differences existed in the strength of these motives. The altruistic scores were highest in Finland, Croatia, New Zealand, the USA, and Canada. "Resume building" was least important in Korea, Japan, and Finland. Learning motives were most important in India, Croatia, and Israel. Social motives received the highest support in the USA, Canada, and Belgium and had the lowest

Table 6.1 Civil society in 13 countries

	Percentage of adults who volunteer[a]	Percentage of students who volunteer	Political regime	Percentage of people who attend religious services at least once a month[b]	Public social expenditure 2003[c]	Individualist ranking (Hofstede, 2001)	Central value types (Schwartz, 1994, 1999)
China	77	84.5	Communist	3.1		20	Hierarchy
USA	66	78.3	Liberal democracy	60.1	16.2	91	Mastery, hierarchy
Canada	47	79.7	Liberal democracy	35.8	17.3	80	Mastery, affective autonomy (English Canada)
Rep. of Korea	47	73.0	Conservative	38.2		18	Hierarchy
UK	43	63.8	Liberal democracy	18.9	20.6	89	Affective autonomy, mastery
Finland	37	68.8	Social democracy	14	22.5	63	Egalitarianism, harmony
Belgium	32	71.9	Corporatist	27	26.5	75	N/A
India	31	86.1	Parliamentary democracy	51.2		48	Hierarchy, conservatism
Netherlands	31	61.6	Corporatist	25.2	20.7	80	Intellectual autonomy, egalitarianism
New Zealand	31[d]	74.5	Liberal democracy	22[e]		79	Mastery, affective autonomy
Croatia	23	50.8	Transitional democracy	52.5		27[f]	N/A (East Europe: harmony, conservatism)
Israel	20	67.6	Corporatist	N/A	12.6	54	Mastery, affective autonomy
Japan	16	39.1	Conservative	12.1	17.7	46	Mastery, affective autonomy

[a]World Values Survey wave 4, 1999/2004
[b]World Values Survey wave 4, 1999/2004
[c]Figures from the OECD Factbook 2008: Economic, Environmental and Social Statistics
[d]Office for the Community and Voluntary Sector, New Zealand, 2007
[e]World Values Survey wave 3, 1994/1999
[f]Score for the former Yugoslavia

Table 6.2 Participation in volunteering and motivations by country

	Participation in volunteering				Motivations in volunteering									
	No		Yes		Resume		Social		Hedonistic		Learning		Altruistic	
USA	247	(21.7)	891	(78.3)	3.71	(0.85)	3.33	(0.83)	2.94	(1.06)	3.74	(0.79)	4.29	(0.95)
Canada	181	(20.3)	711	(79.7)	3.86	(0.97)	3.18	(0.89)	2.87	(1.12)	3.77	(0.80)	4.27	(0.96)
Belgium	248	(28.1)	633	(71.9)	3.22	(0.72)	3.17	(0.62)	2.66	(0.86)	3.50	(0.57)	4.20	(0.60)
China	142	(15.5)	776	(84.5)	3.55	(0.45)	3.06	(0.75)	2.25	(0.98)	3.77	(0.59)	3.99	(0.74)
Croatia	292	(49.2)	302	(50.8)	3.30	(1.04)	2.53	(0.88)	2.13	(1.06)	3.90	(0.72)	4.36	(0.62)
England	209	(36.2)	369	(63.8)	3.69	(0.74)	2.81	(0.81)	2.78	(0.93)	3.78	(0.70)	4.15	(0.86)
Finland	184	(31.2)	405	(68.8)	3.07	(0.81)	2.97	(0.82)	2.43	(0.99)	3.83	(0.60)	4.53	(0.64)
India	83	(13.9)	513	(86.1)	3.25	(0.83)	3.08	(1.00)	3.05	(1.18)	3.99	(0.89)	4.08	(0.99)
Israel	168	(32.4)	351	(67.6)	3.25	(1.11)	2.97	(0.95)	3.01	(1.09)	3.89	(0.68)	4.10	(0.75)
Japan	641	(60.9)	411	(39.1)	3.03	(0.83)	2.73	(0.91)	2.08	(1.01)	3.70	(0.71)	3.93	(0.86)
Korea	187	(27.0)	505	(73.0)	2.70	(0.70)	2.82	(1.75)	3.05	(0.73)	3.44	(0.66)	3.81	(0.72)
Netherlands	227	(38.4)	364	(61.6)	3.38	(0.73)	3.07	(0.63)	2.75	(0.85)	3.46	(0.60)	4.23	(0.62)
New Zealand	146	(25.5)	426	(74.5)	3.48	(0.89)	3.01	(0.82)	2.78	(1.06)	3.74	(0.70)	4.30	(0.84)
Total	2,955	(30.7)	6,657	(69.3)	3.36	(0.95)	3.00	(0.85)	2.65	(1.05)	3.71	(0.71)	4.15	(0.83)

support in Croatia. Protective reasons were most prevalent in India and Korea and are least important in Croatia, Finland, and Japan.

In the analysis on cultural values and volunteering motivation,[1] students who belonged to individualist cultures (scores in Table 6.1) had a statistically significantly higher motivation in resume building. Also, students who belonged to egalitarian cultures (identified in Table 6.1) rated altruistic motivation statistically significantly higher than students from countries with other cultural values. Thus, students from countries whose culture emphasizes voluntary commitment to promoting the welfare of others (egalitarian cultural values of equality, social justice, freedom, responsibility, honesty) rated altruistic motives more important for their volunteering than students from countries with different cultural values. Also, students from countries that have high individualism scores—meaning that the ties between individuals and groups are loose—rated resume motives higher when compared to those from countries with lower individualism score. Effects of conservative culture on social motivation, affective autonomy on protective motivation, and intellectual motive on learning motivation were not statistically significant. In general, female respondents rated altruistic volunteering motives statistically significantly more important in the whole data than male respondents.

6.4 Discussion and Conclusion

Cultural values form the basis for cultural norms and structures, and individuals in cultures know what is good, right, and appropriate in different situations based on these norms (Schwartz, 1994). Individualism and egalitarianism are cultural features which seem to direct the volunteering motives of individuals or the way different motives are acknowledged or explicated among individuals living in countries with such cultural features. It is noteworthy that altruistic motivation was rated central also in countries with dominating cultural values other than egalitarianism. Nevertheless, egalitarian cultural values seem to foster altruistic volunteering motivation or at least encourage individuals to acknowledge and explicate altruistic motivation. Similarly, individualistic cultures seem to foster self-enhancing motivation such as career motivation or encourage individuals to acknowledge and explicate such self-enhancing motives for volunteering.

The influence of cultural values is transmitted through informal and formal socialization. One mediating structure is the political system, which represents and mediates the cultural norms and values of its contexts. It is noteworthy that all countries in the above-summarized study with high scores on individualism (Belgium, Canada, the Netherlands, New Zealand, the UK, and the USA) represent liberal democratic or corporatist political regimes. Also, educational systems mediate

[1] Several factors such as age, gender, family income, and program of study were taken into account, and their influence controlled in the study. For details, see Grönlund et al. (2011).

cultural norms and values, which obviously affect students. For example, the educational system may highlight the career-related benefits of volunteering in these contexts, whereas educational systems in different cultural contexts can be assumed to have other emphases. Countries in the study with egalitarian cultural values (Finland, the Netherlands) have relatively strong public welfare services but different political models (social democratic and conservative). Cross-national comparative studies have shown that countries with strong public welfare have higher scores in social capital—that is, they have generalized trust (in other people), and high levels of participation in charity, political, and other organizations (e.g., Kääriäinen & Lehtonen, 2006). Strong social security can be associated with solidarity and the citizens' trust in the political system (Kumlin & Rothstein, 2003). Egalitarian cultural values can explain both society-level solidarity (public welfare) and individual-level altruistic motivation and behavior (i.e., volunteering). Welfare models are also a way of socializing individuals into cultural values, and the individuals then reinforce these cultural values with their behavior.

On the other hand, there is no support for the hypotheses in the above-summarized study on the connections between other cultural values (except for egalitarianism and individualism). This may be the result of challenges in formulating hypotheses on the connections between the macro-level theoretical approaches and motivation of individuals. The lack of research of this type adds to this difficulty but also highlights the importance of such effort. The study also pointed out some other challenges of cross-cultural research, which have limited previous cross-national studies of volunteering. Both Hofstede (2001) and Schwartz (1994, 2007) discuss the factors involved in conducting cross-national studies. In such research, both the sample and the study instrument must be comparable. The research instrument must be translated into different languages without a loss of meaning. This step involves not only a language translation but also a cultural translation. Slight variations were needed in the above-summarized study even for the English language questionnaires used in Canada, India, New Zealand, the USA, and the UK due to cultural variations and terminology.

Several factors such as age, gender, family income, and program of study were taken into account and their influence controlled in the study summarized above. Despite this, there are several other factors which can influence and explain the connections (or lack of connections) that were found between cultural values and volunteering motivation. The nuanced nature and individual understanding on volunteering motivation is one such factor. Also, as Rochester and colleagues (2010) have pointed out, different values systems coexist in plural societies and provide different foundations for volunteering within a certain country. We must also acknowledge the challenges in testing macro-level influences on individuals' motivations. It is encouraging that individualism and egalitarianism were connected with hypothesized motivation in the research despite these challenges and limitations. Cultural values offer a new context for examining and understanding cross-cultural differences in volunteering. We encourage further research on cultural values in relation to volunteering and also other fields of activity that can have altruistic components, especially based on other cross-nationally comparable samples.

Acknowledgments This chapter is to a large extent based on the following article and the work of the research team that coauthored the article: Grönlund, H., Holmes, K., Kang, C., Cnaan, R., Handy, F., Brudney, J., Haski-Leventhal, D., Hustinx, L., Kassam, M., Meijs, L. C. P. M., Pessi, A. B., Ranade, B., Smith, K. A., Yamauchi, N., and Zrinščak, S. (2011). Cultural values and volunteering: A cross-cultural comparison of students' motivation to volunteer in 13 countries. *Journal of Academic Ethics 9(2)*, 87–106.

References

Anheier, H. K., & Salamon, L. M. (1999). Volunteering in cross-national perspective: Initial comparisons. *Journal of Law and Contemporary Problems, 62*(4), 43–65.

Bekkers, R. (2005). Participation in voluntary associations. Relations with resources, personality, and political values. *Political Psychology, 26*(3), 439–454.

Billis, D. (1993). *Organizing public and voluntary agencies*. London: Routledge.

Carson, E. D. (1999). On defining and measuring volunteering in the United States and abroad. *Journal of Law and Contemporary Problems, 62*(4), 67–71.

Clary, E. G., Snyder, M., Ridge, R. D., Copeland, J., Stukas, A. A., Haugen, J., & Miene, P. (1998). Understanding and assessing the motivations of volunteers: A functional approach. *Journal of Personality and Social Psychology, 74*, 1516–1530.

Cnaan, R. A., & Goldberg-Glenn, R. S. (1991). Measuring motivation to volunteer in human services. *The Journal of Applied Behavioral Science, 27*, 269–284.

Cnaan, R. A., Handy, F., & Wadsworth, M. (1996). Defining who is a volunteer: Conceptual and empirical considerations. *Nonprofit and Voluntary Sector Quarterly, 25*, 364–383.

Cnaan, R., Pessi, A. B., Handy, F., Brudney, J. L., Grönlund, H., Haski-Leventhal, D., Holmes, K., Hustinx, L., Kassam, M., Pessi, A. B., Zrinščak, S., Cnaan, R. A., Yamauchi, N., Smith, K., & Meijs, L. C. P. M. (2012). Student Values, Religiosity, and Pro-Social Behavior: A Cross-National Perspective. Diaconia. *The Journal for the Study of Christian Social Practice, 3*(1), 2–25.

Curtis, J. E., Baer, D. E., & Grabb, E. G. (2001). Nations of joiners: Notions of voluntary association membership in democratic societies. *American Sociological Review, 66*(6), 783–805.

Dekker, P. (2002). On the prospects of volunteering in civil society. *Voluntary Action, 4*(3), 31–48.

Dekker, P., & Halman, L. (2003). Volunteering and values: An introduction. In P. Dekker & L. Halman (Eds.), *The values of volunteering: Cross-cultural perspectives* (pp. 1–6). New York: Kluwer.

Gibbons, J. L. (2013). Guatemalan adolescents' reports of helping in urban and rural Mayan .

Grönlund, H. (2011) Identity and volunteering intertwined: Reflections from the values of young adults. *Voluntas, International Journal of Voluntary & Nonprofit Organizations, 22*(3), 852–874.

Grönlund, H., Holmes, K., Kang, C., Cnaan, R., Handy, F., Brudney, J., ... Zrinščak, S. (2011). Cultural values and volunteering: A cross-cultural comparison of students motivation to volunteer in 13 countries. *Journal of Academic Ethics 9*(2), 87–106.

Gupta, A. G. (2013). Altruism in Indian religions: Embracing the biosphere. In D. A. Vakoch (Ed.), *Altruism in cross-cultural perspective*. New York: Springer.

Halman, L. (2003). Volunteering, democracy and democratic attitudes. In P. Dekker & L. Halman (Eds.), *The values of volunteering: Cross-cultural perspectives* (pp. 179–198). New York: Kluwer.

Handy, F., Cnaan, R. A., Brudney, J., Meijs, L., Ascoli, U., & Ranade, S. (2000). Public perception of "who is a volunteer": An examination of the net-cost approach: A cross-cultural perspective. *Voluntas, 11*(1), 45–65.

Haski-Leventhal, D. (2009). Altruism and volunteerism: The perceptions of altruism in four disciplines and their impact on the study of volunteerism. *Journal for the Theory of Social Behaviour, 39*(3), 271–299.

Hodgkinson, V. (2003). Volunteering in global perspective. In P. Dekker & L. Halman (Eds.), *The values of volunteering: Cross-cultural perspectives* (pp. 35–53). New York: Kluwer.
Hofstede, G. (2001). *Culture's consequences: Comparing values, behaviors, institutions and organizations across nations*. Thousand Oaks, CA: Sage.
Inglehart, R. (1997). *Modernization and postmodernization: Cultural, economic, and political change in 43 societies*. Princeton: Princeton University Press.
Inglehart, R. (2003). Modernization and volunteering. In P. Dekker & L. Halman (Eds.), *The values of volunteering: Cross-cultural perspectives* (pp. 55–70). New York: Kluwer.
Kääriäinen, J., & Lehtonen, H. (2006). The variety of social capital in welfare state regimes – a comparative study of 21 countries. *European Societies, 8*(1), 27–57.
Kumlin, S., & Rothstein, B. (2003). *Investing in social capital: The impact of welfare state institutions*. Paper prepared for delivery at the 2003 Annual Meeting of the American Political Science Association, August 31, 2003. Retrieved from http://www.students.uni-mainz.de/bonea001/Dokumente/KumlinRothsteinAPSA2003.pdf.
Musick, M. A., & Wilson, J. (2008). *Volunteers: A social profile*. Bloomington, IN: Indiana University Press.
Musick, M. A., Wilson, J., & Bynum, W. B. (2000). Race and formal volunteering: The differential effects of class and religion. *Social Forces, 78*(4), 1539–1570.
Office for the Community and Voluntary Sector. (2007). *Facts and figures on volunteering in New Zealand*. Wellington: Author.
Organisation for Economic Co-operation and Development (OECD). (2008). *OECD factbook 2008: Economic, environmental and social statistics 2008*. Paris: Author.
Parboteeah, K. P., Cullen, J., & Lim, L. (2004). Formal volunteering: A cross national test. *Journal of World Business, 39*, 431–441.
Rochester, C., Paine, A. E., Howlett, S., & Zimmeck, M. (2010). *Volunteering and society in the 21st century*. Hampshire: Palgrave Macmillan.
Rokeach, M. (1973). *The nature of human values*. New York: Free Press.
Ruiter, S., & De Graaf, N. D. (2006). National context, religiosity, and volunteering: Results from 53 countries. *American Sociological Review, 71*(April), 191–210.
Salamon, L. M., & Anheier, H. K. (1998). Social origins of civil society: Explaining the nonprofit sector cross-nationally. *Voluntas, 9*(3), 213–248.
Salamon, L. M., & Sokolowski, S. W. (2003). Institutional routes of volunteering: Toward a macro-structural theory of individual voluntary action. In P. Dekker & L. Halman (Eds.), *The values of volunteering: Cross-cultural perspectives* (pp. 71–90). New York: Kluwer.
Schimmack, U., Oishi, S., & Diener, E. (2005). Individualism: A valid and important dimension of cultural differences between nations. *Personality and Social Psychology Review, 9*(1), 17–31.
Schwartz, S. H. (1992). Universals in the content and structure of values: Theoretical advances and empirical tests in 20 countries. Teoksessa M. P. Zanna (Toim.). *Advances in experimental social psychology, 25*, 1–65.
Schwartz, S. H. (1994). Beyond individualism and collectivism: New cultural dimensions of values. In U. Kim, H. C. Triandis, C. Kagitcibasi, S.-C. Choi, & G. Yom (Eds.), *Individualism and collectivism: Theory, method and application* (pp. 85–119). Thousand Oaks, CA: Sage.
Schwartz, S. H. (1999). A theory of cultural values and some implications for work. *Applied Psychology: An International Review, 48*(1), 23–47.
Schwartz, S. H. (2007). Universalism values and the inclusiveness of our moral universe. *Journal of Cross-Cultural Psychology, 36*, 711–728.
Schwartz, S. H. (2012). *Cultural value orientations: Nature and implications of national differences*. Retrieved from http://www.tau.ac.il/law/cegla3/tax2/1%20Cultural%20Value%20Orientations%20-%20Schwartz.pdf.
Smith, S. W., Lapinski, M. K., Bresnahan, M. J., & Smith, S. L. (2013). Conceptual aspects of altruism in cross-cultural contexts. In D. A. Vakoch (Ed.), *Altruism in cross-cultural perspective*. New York: Springer.

Stadelman-Steffen, I., & Freitag, M. (2011). Making civil society work: Models of democracy and their impact on civic engagement. *Nonprofit and Voluntary Sector Quarterly, 40*(3), 526–551.
2011 State of the World's Volunteerism Report: Universal values for global well-being. Retrieved from: http://www.unric.org/en/images/stories/2011/PDF/SWVR%20Report%20[Eng].pdf.
Wuthnow, R. (1991). *Acts of compassion*. Princeton: Princeton University Press.
Yeung, A. B. (2004). An intricate triangle: Religiosity, volunteering, social capital: European perspective – the case of Finland. *Nonprofit and Voluntary Sector Quarterly, 33*(3), 401–423.
Ziemeck, S. (2003). *The economics of volunteer labour supply: An application to countries of a different development level*. New York: Peter Lang.

Chapter 7
Daoism and Altruism: A China–USA Perspective

Yueh-Ting Lee, Wenting Chen, and Sydney Xinni Chan

7.1 Introduction

Chinese Daoist beliefs are philosophically and psychologically complicated and far reaching (Lee, Yang, & Wang, 2009). One of the fundamental questions raised by Chinese Daoism Is: "Where did its ideas come from?" To answer this question, we should first look at shamanism. Over 8,000 years ago, fundamental religious belief in China was a form of shamanism (Lee & Li, 2011; Lee & Wang, 2003; Xu, 1991; Yuan, 1988), a spiritual belief or practice in which a shaman can connect the inner world with the outer world, the body with the soul, and the living with the dead (Krippner, 2002; Meng, 2005; Seyin, 1998). As time went on, Daoism, Confucianism, and other fundamental Chinese religious and philosophical ideologies developed out of shamanism and affected Chinese behavior and thinking on an almost daily basis for thousands of years (see Chen & Lee, 2008; Hsu, 1981; Lee & Wang, 2003). In this chapter, we focus only on Daoism and altruism.

Because Chinese Daoist belief is closely related to altruism, we first address basic issues related to Laozi's Daoism and philosophical aspects of human altruism. Although Daoism originated in China, its impact seems universal with regard to human beliefs and behaviors (see Lee & Haught, 2012). Our research indicates that there has been much research on Daoism and a moral sense of altruism (Chen, 2009; Hu & Ma, 2000; Xu & Lu, 2012), but little attention has been paid to Daoist

Y.-T. Lee (✉) • S.X. Chan
Department of Psychology, University of Toledo, 2801 West Bancroft Street, Toledo, OH 43606-3390, USA
e-mail: YT.Lee@Utoledo.edu; sydneychan82@gmail.com

W. Chen
Department of Curriculum & Instruction, University of Toledo, 2801 West Bancroft Street, Toledo, OH 43606, USA
e-mail: wentingchenphd@gmail.com

D.A. Vakoch (ed.), *Altruism in Cross-Cultural Perspective*,
International and Cultural Psychology, DOI 10.1007/978-1-4614-6952-0_7,

altruism, except Kirkland's work, which dealt with altruism from the perspective of Daoist religion (Kirkland, 1986, 2002). Our chapter focuses on Daoism and altruism psychologically and philosophically. First, we will discuss the controversial debate on the nature of altruism from both Western and Eastern perspectives. Next, we discuss the Daoist water-like Big-Five model (which includes altruism) and review at great length the empirical results based on this model. Finally, the chapter concludes with a look at the implications of and prospects for this research.

7.2 Basic Controversies and Findings in the East and West: Altruism and Daoism

7.2.1 Are Human Beings Altruistic? What Do We Mean by Altruism?

In the eyes of certain people, human beings are basically kind and altruistic by nature. However, others believe that humans are selfish by nature. There has been a debate among Westerners on this issue (see Post, 2003, pp. 59–61) ever since Plato, who emphasized the primacy of reason in moral motivation, ethics, and/or morality, including altruism. Altruism serves as a lightning rod. Though scholars such as Thomas Hobbes, Friedrich Nietzsche, Sigmund Freud, Edwin O. Wilson, and Richard Dawkins saw human beings as selfish, others (e.g., Adam Smith, David Hume) tended to believe that altruistic motives do exist within the repertoire of human nature. Immanuel Kant was different from everyone named above because he had a neutral perspective on altruism (see Nagel, 1970) and argued that certain altruistic inclinations exist but are not to be trusted because they are unstable and unreliable. To Kant, altruistic behavior is possible but must be grounded in a categorical rational imperative (Nagel).

There was a similar debate on altruism and human nature among scholars in ancient China. For Mencius (371–289 BC), human beings are basically good and altruistic. Everyone should unconditionally do what he or she ought to do and "extend himself or herself so as to include others" (Fung, 1948, p. 69). Basically, people cannot bear to see the suffering of others. For example, if someone sees a child about to fall into a well, that person will without exception experience a feeling of alarm and distress. Trying to rescue the child is an example of altruistic behavior or human goodness.

On the other hand, Xunzi, who lived at the same time of Mencius, held that the nature of man is evil and selfish. People are born with an inherent desire for profit and sensual pleasure (Fung, 1948, p. 145). However, in Kaozi's view (Fung), human beings can be either selfish or unselfish. Kaozi's argument was somewhat similar to Kant's.

In brief, altruism as part of human nature is controversial to both Westerners and Easterners. Though we cannot focus too much on human nature, altruism is worth further investigation. What is altruism? How do social scientists define it?

Howard and Piliavin (2000) defined altruism as helping others in the absence of psychological rewards that benefit the agents. In psychology, they also held that altruistic behavior should exclude any material motive (Post, 2003, p. 59). Harvard sociologist Sorokin (2002/1954) defined altruism as the five-dimensional universe of psychosocial love (i.e., intensity, extensity, duration, purity, and adequacy). For example, the intensity of love (or altruistic behavior) tends to decrease with an increase of duration. The intensity, purity, and adequacy of love are somewhat more frequently associated positively than negatively or not at all. Adequate love is likely to last longer than inadequate love (also see Post, 2003).

In social psychology, Moghaddam (1998) defined altruism as a "behavior intended to help another, without regard for benefit to oneself" (p. 297). Batson (1991) defined altruism as a motivational state with the ultimate goal of increasing another's welfare. Batson also pointed out that altruism does not necessarily involve self-sacrifice, although it is inherently self-sacrificial. Thus, we argue that altruism is more effective if it involves self-sacrifice, which is consistent with Chinese Daoism/Taoism. This brings us to the relationship between culture, altruism, and Daoism/Taoism.

7.2.2 *Cultural Research on Self-sacrificing Altruism*

There is a difference between Easterners and Westerners with regard to generosity and unselfishness, which are implicitly related to altruism (Lee & Seligman, 1997; Tang, Furnham, & Davis, 2002; Yik & Bond, 1993; Zhang, Lee, Liu, & McCauley, 1999). For example, in developmental psychology, Ma and her colleagues (Ma, 1992, 2003; Ma & Leung, 1991) employed the Child Altruism Inventory which they developed and found that Hong Kong Chinese children's altruistic behavior was directly related to a positive family environment and positive peer influence.

From a perspective of self-sacrifice, these researchers found that people with high moral judgment were more willing than people with low moral judgment to sacrifice their lives for any recipient and to rescue a stranger (Ma, 1992). Further, by interviewing 37 Chinese in Hong Kong who had donated bone marrow to an unrelated recipient, Holroyd and Molassioitis (2000) found that this type of Chinese altruism or donation was more of a self-fulfilling act (i.e., yielding self-satisfaction or self-growth) than a social act, with very little familial or social recognition being accorded in the public world, which is totally different from Western culture.

7.2.3 *Connections Between Taoism/Daoism and Altruism:* **Dao, De,** *and* **Shui** *(or Water)*

Lee (2003) noted that Laozi, who has been recognized as the founder of Daoism (or Taoism), contended that the best qualities or personalities are like water because all species and organisms depend on water (which is to be discussed in greater length below). These "wateristic" personality attributes, Lee argues, affect Chinese notions of altruism. In traditional Chinese beliefs, people with a good Taoist personality should be as altruistic as water. Philosophically, water is modest and humble. It always goes to the lowest place. Since it always remains in the lowest position and does not compete, it is not only helpful and beneficial to all things but also implies self-sacrifice, which is psychologically very satisfying to any Taoist.

Further, we can see connections between Taoism and altruism when examining Laozi's *Dao* and *De* (Sima, 1994). *Dao* can mean a road, a path, the way it is, the way of nature, the way of ultimate reality, and the rules and laws of nature. According to Blankney (1955), in the eyes of Chinese, *Dao* refers not only to the way the whole world of nature operates but also signifies the original undifferentiated Reality from which the universe evolved. *De* means humanistic behavior/virtues, character, influence, or moral force. The character *De* consists of three parts: (1) an ideographic meaning "to go"; (2) another, meaning "straight"; and (3) a pictograph meaning "the heart." Put together, these imply motivation by inward rectitude (Blakney, p. 38).

In another translation (see Addiss & Lombardo, 1993), Dao means a "way" in both the literal ("road") and metaphysical ("spiritual path") sense. More rarely, it can also mean "to say," "to express," or "to tell." According to Burton Watson (see Addiss & Lombardo, p. xiii), Dao literally is a "way" or "path" and is used by other schools of Chinese philosophy to refer to a particular calling or mode of conduct. However, in Daoistic writing, it has a far more comprehensive meaning, referring rather to a metaphysical first principle that embraces and underlies all beings, a vast Oneness that precedes and in some mysterious manner generates the endlessly diverse forms of the world. Thus, *Dao* lies beyond the power of language to describe. Burton Watson (see Addiss & Lombardo, p. xiii) defined *De* as the moral virtue or power that one acquires through being in accord with the *Dao*, what one gets from *Dao*.

The best is like water (*shang shan ruo shui*)—this is what we call the Daoist/Taoist model of "wateristic" or water-like personality (Lee, 2003, 2004; Watts, 1975) which includes five essential components: (1) altruism, (2) modesty/humility (or humbleness), (3) flexibility, (4) transparency and honesty, and (5) gentleness with perseverance. These are discussed in greater detail in the following sections.

7.3 Research on the Daoist Water-Like Big-Five Model and Altruism: A Theoretical and Empirical Perspective

7.3.1 Daoism and the Water-Like Daoist Big-Five Model with Altruism

One of the most important components of Daoist beliefs can be summarized as "the best is like water" (i.e., "*Shang Shan Ruo Shui*" in Chinese) by Laozi. In other words, a great individual must act like water in five ways. Based on the narrative and exploratory discussion by Lee and colleagues (see Lee, Han, Bryron, & Fan, 2008; Lee & Haught, 2013; Lee, Yang, et al., 2009), Daoist Big-Five or water-like personality is summarized as follows.

7.3.1.1 Water Is Altruistic

All species and organisms depend on water. Without water, none of them can survive. What does water get from us? It gets almost nothing. Accordingly, Daoist individuals should be as altruistic as water, which is very helpful and beneficial to all things.

The highest value (or the best) is like water,
The value in water benefits All Things
And yet it does not contend,
It stays in places that others despise,
And therefore is close to *Dao*. (*Laozi*: Chap. 8; see Wing, 1986)

Daoism recognizes that the ultimate goal of sages or cultivated individuals is to serve their people without the desire for personal benefit or gratitude. Laozi stated that, "The best are like water, good at benefiting all things without competing for gaining" (*Laozi*: Chap. 8). This entails selflessness as an essential attribute of a sage, which is realized by accepting other people's aspirations as one's own. "The sage does not have aspirations but adopts those of the people as his own" (*Laozi*: Chap. 49). Only when an individual does not have his own ambitions can he truly serve his people instead of competing with them.

7.3.1.2 Water Is Very Modest and Humble

Do we not always see water go to the lowest place? As we can see from the above quotation (i.e., *Laozi*: Chap. 8), although water benefits all things, it does not contend with anything and always stays in the lowest places that others despise. While many Westerners often value and enjoy a sense of authority, assertiveness, aggressiveness, and competitiveness, Laozi encouraged people to have a water-like characteristic—that is, to maintain a low profile, to be humble and modest, especially in the face of the Dao or nature, and to be very helpful and/or beneficial to others.

To Laozi, modesty or humbleness, willingness to help and benefit others, and the ability to maintain a low profile (just like water) are qualities essential to an individual or sage who wants to influence others:

The rivers and seas lead the hundred streams
Because they are skillful at staying low.
Thus they are able to lead the hundred streams. (*Laozi*: Chap. 66)

In Laozi's opinion, people who are humble and modest not only exist in good harmony with others but are also effective leaders, just like the rivers and seas.

The sea, for instance, can govern a hundred rivers because it has mastered being lower. Being humble is important for individuals because it enables them to accept people's goals as their own and to attract and unite people around themselves. Laozi said, "He/She who knows how to motivate people acts humble. This is the virtue of no rival and uses the strength of others" (*Laozi*: Chap. 68).

7.3.1.3 Water Is Very Adaptable and Flexible

It can stay in a container of any shape. This flexibility and fluidity lend it a great deal of wisdom to influence others. Good individuals or sages can adjust themselves to any environment and situation just as water does to any container. Maintaining flexibility and adapting to the dynamics of change, like water following its path, are probably the best options for a leader.

7.3.1.4 Water Is Transparent and Clear

People should be honest and transparent to those around them. The most honorable individuals (not only leaders) are usually honest and transparent like water. Though Western Machiavellian or other deceptive approaches might work temporarily, being honest and transparent is one of the big ethical concerns in modern management. Water itself is very clear and transparent if you do not make it muddy. In Chap. 15, Laozi stated, "Who can (make) the muddy water clear? Let it be still, and it will gradually become clear." Metaphorically, human beings by nature are naive and honest. Social environment and competition (like muddiness) make them unclear. Water's clarity, transparency, and honesty are most appreciated by Laozi.

7.3.1.5 Water Is Very Soft and Gentle But Also Very Persistent and Powerful

If drops of water keep pounding at a rock for years, even the hardest rock will yield. Over time, water can cut through the hardest rock, forming valleys and canyons. The style of sages or individuals should be similarly gentle and soft but persistent and powerful. Here is an example of what we could learn from water:

Nothing in the world
Is as yielding and receptive as water;
Yet in attacking the firm and inflexible,
Nothing triumphs so well. (*Laozi*: Chap. 78)

Because there is nothing softer than water, yet nothing better for attacking hard and strong things, there is no substitute for it. Its softness enables it to tolerate all kinds of environments, gathering strength without wearing out at an early stage. The resolution and perseverance of water help it to cut a path through hard rocks and wear away mountains. It is very important for a leader to know this dialectical relationship and to acquire the resolute and persevering characteristics of water.

7.3.1.6 Summary

Water has five features that are essential to all individuals, from sages to regular people. This is what we call the Daoist/Taoist model of "wateristic" personality

(Lee, 2003, 2004; Lee, Norasakkunkit, Li, Zhang, & Zhou, 2008; Watts, 1961, 1975) which includes five essential components: (1) altruism, (2) modesty/humility (or humbleness), (3) flexibility, (4) transparency and honesty, and (5) gentleness with perseverance (Lee, 2003, 2004; Lee & Haught, 2013; Lee, Norasakkunkit, et al., 2008). This model is summarized in Fig. 7.1.

Altruism is one of the most essential components in this model. We now review certain empirical data that are related to altruism in the Daoist Big-Five model.

7.3.2 Empirical Results of Daoist Altruism Across Cultures

In the following section, we review how altruism is seen in two empirical USA–China studies of the Daoist Big-Five model. The first study was reported by Lee, Norasakkunkit et al. (2008). The second one was reported in 2012 (Lee & Haught, 2013).

7.3.2.1 Study 1: Comparing Chinese and American Altruism

A total of 228 college students participated in this study. There were 122 Chinese students from a university in North China and 106 students from a university in the Midwest. One hundred and forty-one participants were women, and the mean age of the group was 20.39 years.

To measure altruism, we create various scenarios (helping a person with a chronic disease, or HIV/AIDS, a war victim, and a victim of catastrophe) as follows:

Scenario 1: Suppose one day you see a person who cannot move and is lying on the ground. You are informed that the person has a chronic disease (e.g., hepatitis or tuberculosis) which is probably contagious. The person pleads for help.

Scenario 2: Suppose one day you see a person who has collapsed and is lying on the ground. You are informed that the person has HIV/AIDS which is probably contagious. The person pleads for help.

Scenario 3: Human beings sometimes cannot avoid conflict or violence. War is part of human conflict or violence. Suppose one day you are in a situation where you see a group of people fighting or killing another group. Though you do not know which group is the aggressor or which one is the victim, you notice many people injured or killed. If you plan to rescue or help one of the victims, you may bring danger to yourself (injury or death). One person pleads for help.

Scenario 4: Human beings are sometimes so vulnerable and weak that they cannot control natural disasters or catastrophes (earthquake, flood, or fire). Suppose one day you encounter victims of such disasters and one of the victims pleads for help.

All participants were informed that the purpose of this study was to investigate "how individuals make personal decisions when facing various challenging situations" and were asked to answer each question based on the Likert scale from 1 (least likely) to 7 (most likely) with regard to the five issues that make up our

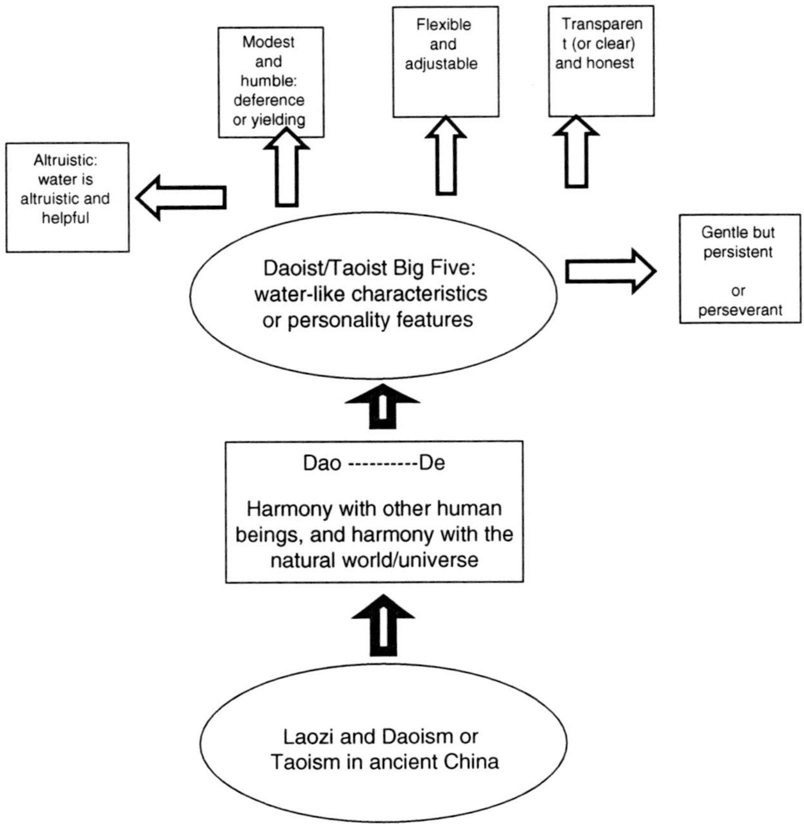

Fig. 7.1 The Daoist/Taoist model of wateristic personality (Taoist Big Five)

dependent measures (helping at the cost of one's health, helping at the cost of one's life, blood donation, time donation, and money donation):

How likely are you to rescue/help the person directly if it were to put your *health* at risk?
How likely are you to rescue/help the person directly if it were to put your *life* at risk?
How likely are you to donate your blood to that person if it is called for?
How likely are you to donate your time (2 weeks) to the person if it is called for?
How likely are you to donate your money (about $500) to that person if it is called for?

(A note about the value of money: Instead of $500, Chinese participants were asked to donate 2,000 Renminbi [or 2,000 Yuan], which is psychologically and culturally equivalent to the dollar amount but not economically.)

Additionally, we also used a scenario that involved encountering aliens. "Suppose 1 day you encounter a group of aliens. You do not understand what they are talking about and they do not understand what you are trying to say. From your intuitive

Table 7.1 Mean (*SD*) altruism toward various types of "victims" as a function of culture

	USA	China	
	(*N*=106)	(*N*=122)	*F*-value (*df*)
a. Patients with chronic diseases	4.69 (1.24)	3.70 (1.21)	36.67 (1,226)***
b. HIV and AIDS patients	4.61 (1.35)	3.52(1.56)	31.13 (1,226)***
c. War victims	4.44 (1.40)	3.44(1.41)	31.22 (1,226)***
d. Catastrophe victim	5.28 (1.28)	4.37 (1.23)	29.58 (1,226)***

Greater numbers mean more altruism
**$p<0.01$; ***$p<0.001$

Table 7.2 Mean (*SD*) altruism toward alleged victims across all situations as a function of culture

	USA	China	
	(*N*=106)	(*N*=122)	*F*-value
a. Help with health risk	4.74 (1.43)	3.68 (1.61)	26.91 (1,226)***
b. Help with life risk	4.12 (1.58)	2.71(1.48)	48.16(1,226)***
c. Donating blood	5.99 (1.33)	4.91(1.73)	27.54(1,226)***
d. Donating time	4.94 (1.46)	3.84 (1.55)	29.74(1,226)***
e. Donating money	3.75 (1.77)	3.41 (1.84)	1.91(1,226) ns

Greater numbers mean more altruism
**$p<0.01$; ***$p<0.001$

judgment, one of them seems to plea for help." On the scale from 1 (least likely) to 7 (most likely), participants were asked about the following questions:

How likely are you to help that alien?
How likely are you to feel frightened of the aliens?
How likely are you to escape from the situation?
How likely are you just to observe and not take any action?
How likely are you to blow those aliens away if you have a power weapon?

Lee, Norasakkunkit et al. (2008) found that Americans were more altruistic than Chinese. With respect to altruism or self-sacrificial altruism, the reliability scores for dependent measures in Scenario 1 through 4 ranged from alpha=0.76 to alpha=0.82. Americans were found to be more altruistic across all four situations, such as helping patients with chronic diseases, with HIV/AIDS, and the victims of war and catastrophe (see Table 7.1).

With regard to providing help that involves health risk, life risk, and blood and time donation, Americans were more altruistic than Chinese with the exception of money donation (where there was no significant difference)—see Table 7.2. This is also true with situations that involve helping victims of natural disasters or catastrophe (see Table 7.3), which may help us to better understand responses to the tsunami tragedy in South Asia.

Consistently, our regression analysis revealed that culture played an important role in altruistic behavior (Beta=0.40, $t=6.46$, $p<0.001$). In other words, overall Americans were more altruistic than Chinese.

Table 7.3 Mean (*SD*) altruism toward an assumed victim in a hypothetical natural disaster as a function of culture

	USA	China	
	(N=106)	(N=122)	F-value
a. Help with health risk	5.50 (1.51)	4.31 (1.78)	29.08***
b. Help with life risk	5.02 (1.80)	3.42(1.90)	42.24***
c. Donating blood	6.19 (1.37)	5.41(1.70)	14.20***
d. Donating time	5.39 (1.76)	4.65 (1.81)	10.41**
e. Donating money	4.28 (2.05)	4.07 (2.10)	0.58 ns
Total	5.28 (1.28)	4.37 (1.23)	29.58***

Greater numbers mean more altruism
**$p<0.01$; ***$p<0.001$

Table 7.4 Mean (*SD*) attitudes toward aliens in various situations as a function of culture

	USA	China	
	(N=106)	(N=122)	F-value
a. Help the alien	4.25 (2.06)	5.16 (1.74)	13.03 (1,226)***
b. Feel frightened	5.00 (2.01)	4.06 (1.72)	14.59 (1,226)***
c. Escape	4.14 (1.85)	3.17 (1.59)	18.08 (1,226)***
d. No action	3.71 (1.93)	3.53 (1.68)	0.54 (1,226) ns
e. Blow them away	2.80 (2.10)	1.76 (1.12)	22.60 (1,226)***

Greater numbers mean more likelihood of behavior
**$p<0.01$; ***$p<0.001$

However, Lee, Norasakkunkit, et al. (2008) found that Chinese seemed to be more sympathetic than Americans toward aliens. When they are asked to imagine encountering aliens, this measure focuses on people's willingness to be altruistic toward out-group members whose intentions are uncertain. As can be seen in Table 7.4, Chinese participants (M=5.16) tended to be more willing to help aliens than their American counterparts (M=4.25), F (1,226)=13.03. There were no significant differences regarding seeing aliens and taking no action. But Chinese people reported feeling less frightened by the aliens and were less likely to destroy them than their American participants. In other words, when compared with Chinese, Americans tended to act more aggressively toward out-group members whom they do not know and encounter in uncertain situations (see Table 7.4).

7.3.2.2 Study 2: How Was Daoist Altruism Measured Scientifically?

As described by Lee and Haught (2013), a total of 261 students from a US Midwestern research university participated in this study online via Psychdata, an online data collection platform. We had 96 males and 165 females with a mean age of 20.01 (SD=3.61). A majority of them were European–American students with Christian–Catholic background (N=211). The remainder practiced other religions.

Table 7.5 Correlations between Daoist Big Five: Water-like leadership styles

Daoist Big Five	1	2	3	4	5
1. Altruism	–				
2. Modesty	0.40**	–			
3. Flexibility	0.12*	0.11	–		
4. Honesty	0.09	0.24**	0.07	–	
5. Gentleness and perseverance	0.46**	0.26**	0.25**	0.08	–

*$p<0.05$; **$p<0.01$

Questions that measured the Daoist Big-Five water-like leadership style were based primarily on the HEXACO measure developed by K. Lee and his colleagues (Ashton & Lee, 2008; Lee & Ashton, 2004; Lee, Ashton, Morrison, Cordery, & Dunlup, 2008; Lee, Ashton, et al., 2009) with regard to altruism, modesty, flexibility, honesty, and gentleness. Because the HEXACO scale did not measure perseverance or persistence, we added the GRIT scale (see Duckworth, Peterson, Mathews, & Kelly, 2007) to our measure. All the questions extracted from HEXACO and GRIT scales were measured on a five-point Likert scale from 1 (strongly disagree) to 5 (strongly agree). Here is a list of those items (those marked with * were reversed in scoring):

I have sympathy for people who are less fortunate than I am.
I try to give generously to those in need.
It wouldn't bother me to harm someone I didn't like.*
People see me as a hard-hearted person.*

In their study, Lee and Haught (2013) reported that their results had a very satisfactory internal consistency (alpha=0.62). Second, this Daoist altruistic scale was negatively correlated with the Machiavellian scale ($r=-0.43$, $p<0.001$) but positively correlated with the good human nature scale ($r=0.40$, $p<0.001$). Third, they found that altruism was related to each dimension except for honesty (see Table 7.5). Finally, they found that female respondents ($M=3.80$) tended to be more altruistic than male students ($M=3.54$), $t(259)=-2.33$, $p=0.004$, two tailed.

7.4 Implications and Conclusion

The conceptual discussion and results reviewed above show that Chinese Daoism is broader and more meaningful scientifically and intellectually than one might think. As part of human nature, altruism is controversial both in the Eastern and Western cultures. However, we feel that focusing on Daoist altruism and investigating the Daoist water-like Big-Five model may help us to understand and appreciate both Chinese philosophy and human altruism cross-culturally and psychologically. As a religion, Daoism is linked to altruism (see Kirkland, 1986, 2002). However, and here we differ from Kirkland's research, our emphasis is not on its religion. Instead, we investigate Daoism more philosophically, psychologically, and scientifically.

More specifically, in our empirical study of Daoist altruism and the water-like Big-Five model, several cross-cultural findings are worth summarizing and reiterating. First, according to Lee, Norasakkunkit et al. (2008), Americans were found to be more altruistic than Chinese in terms of specific self-sacrificial altruistic behaviors. But the Chinese had a more positive attitude toward aliens than Americans. This may be related to the fact that the post-9/11 environment in the USA could make Americans less willing to trust out-group members whose intentions are uncertain and therefore less willing to help them (Todd, 2003).

Second, Lee and Haught (2013) also recently found that Daoist altruism can be measured scientifically with high consistency and validity. Women were found to be more altruistic than men, which conforms to the stereotype and is consistent with other work (see Eagly & Crowley, 1986; Lee, 1995, 2000, 2011; Lee & Jussim, 2010; Lee & Ottati, 1993; Lee, Jussim, & McCauley, 1995, 2013; Lee & Ottati, 1993). However, more research will be needed before a definitive conclusion is reached.

Why should we study Daoist altruism and Daoism psychologically across cultures? Why are they important theoretically and practically? First, Taoism/Daoism and Laozi's philosophy focus on harmony (see Lee, 2003): (a) being humanistic and harmonious with other humans and (b) being harmonious with Mother Nature (or the universe). Based on Daoism and social sciences, a Taoist (or Daoist) Big-Five model of water-like (or wateristic) personality involving altruism, modesty, flexibility, transparency/honesty, and perseverance provides us with a new approach to understanding human behavior (see Fig. 7.1). Perhaps Daoism/Taoism, including altruism and modesty, could be something we see general across cultures.

Second, the water-like (or wateristic) personality, including altruistic tendencies, may lead to more peace and harmony. Our research could throw light onto solutions to major world problems (Lee, 2003; Lee, Bumgarner, Widner, & Luo, 2007; Lee, McCauley, & Draguns, 1999; Lee, McCauley, Moghaddam, & Worchel, 2004; Lee, Takaki, Ottati, & Yan, 2004). Natural resources cannot continue to be overexploited because oil, for instance, will be gone. An ethnocentric and narcissistic military strategy may work temporarily but only, at best, delay the problem for a while. A bigger conflict between more powerful nations fighting for rights to scarce resources available in small, helpless countries may escalate into another world war. If humans kill each other (with nuclear weapons, other weapons of mass destruction, cyber wars, or any other type of force) and continue damaging the environment, and thus continue overusing or overconsuming resources provided by Mother Nature or earth, can we survive? Originally, Daoism came from shamanism and totemism, which are very naturalistic and humanistic (Chen & Lee, 2008; Lee, Jussim, & McCauley, 1995, 2013; Lee & Wang, 2003; Triandis, 2001). We cannot survive too long if we oppose other humans and work against the natural or external world.

One more interesting question is if a much more powerful alien force were to make contact with us, would we hope that they were altruistic and yielding to our ways, as well as persistent about working with us toward a greater harmony? In brief, Daoist harmony is meaningful and necessary for all species, including aliens or other humans, to coexist.

Finally, consistent with the water-like (or wateristic) personality or the Daoist Big-Five model, other research suggests that leaders who displayed self-sacrificial altruistic behavior lead their followers more effectively than those without self-sacrificial altruism (Choi & Mai-Dalton, 1998, 1999; van Knippenberg & van Knippenberg, 2005). If self-sacrificial altruism is powerful in the field of management and leadership, it can surely play a beneficial role in other domains, such as in unlimited love for human beings (Post, 2003) and altruistic behavior toward children in Hong Kong (see Ma, 2003; Ma & Leung, 1991).

Where shall we go from here with regard to future research directions? First, though this research may shed both theoretical and empirical light on this topic, more research will be needed to test various aspects of the Daoist Big-Five model of wateristic (or water-like) personality (e.g., altruism, modesty, flexibility, honesty, and perseverance). Second, theoretically and cross-culturally, we will need to expound more specifically on how Chinese Daoism is related to totemism and shamanism and how it is related to other world philosophies and religions. Much more work is needed in these areas. Though the research here is far from perfect, "a journey of thousand miles begins with a single step" as is stated in Laozi's Tao De Jing (Chap. 64).

Acknowledgments We extend our thanks to Heather Rasinski, Mark Ratta, Timothy Bodie, and the assistants in our SCIR (Social Cognition and Intercultural Relations) Laboratory for their helpful comments on prior versions of this chapter. Part of the data cited in this chapter was originally reported elsewhere by Lee and colleagues (Lee & Haught, 2013; Lee, Norasakkunkit, et al., 2008).

References

Addiss, S., & Lombardo, S. (1993). *Lao-Tzu Tao Te Ching* (Trans.) (Introduced by Burton Watson). Indianapolis, IN: Hackett Publishing Co.

Ashton, M., & Lee, K. (2008). The prediction of Honesty–Humility-related criteria by the HEXACO and Five-Factor Models of personality. *Journal of Research in Personality, 42*, 1216–1228.

Batson, C. D. (1991). *The altruism question: Toward a social psychological answer*. Hillsdale, NJ: Lawrence Erlbaum Associates.

Blakney, R. B. (1955). *The way of life: Lao Tzu (a new translation of the Tao Te Ching)*. New York: New American Library.

Chen, Y. (2009). Restrictions of ancient Chinese science by Confucianism and Taoism: Also as an answer to Needham's grand question. *Social Science Academic Journal of Tsinghua University, 1*(24), 116–126.

Chen, C. C., & Lee, Y.-T. (2008). *Leadership and management in China: Philosophies, theories and practices*. New York: Cambridge University Press.

Choi, Y., & Mai-Dalton, R. R. (1998). On the leadership function of self-sacrifice. *The Leadership Quarterly, 9*, 475–501.

Choi, Y., & Mai-Dalton, R. R. (1999). The model of followers' responses to self-sacrificial leadership: An empirical test. *The Leadership Quarterly, 10*, 397–421.

Duckworth, A. L., Peterson, C., Matthews, M. D., & Kelly, D. R. (2007). Grit: Perseverance and passion for long-term goals. *Journal of Personality and Social Psychology, 92*(6), 1087–1101.

Eagly, A. H., & Crowley, M. (1986). Gender and helping behavior: A meta-analytic review of the social psychological literature. *Psychological Bulletin, 100*, 283–308.
Fung, Y.-L. (1948). *A short history of Chinese philosophy*. New York: Macmillan.
Holroyd, E., & Molassiotis, A. (2000). Hong Kong Chinese perceptions of the experience of unrelated bone marrow donation. *Social Sciences and Medicine, 51*(1), 29–40.
Howard, J. A., & Piliavin, J. A. (2000). Altruism. In E. F. Borgatta & R. J. V. Montgomery (Eds.), *Encyclopedia of sociology* (p. 114). New York: Macmillan Reference.
Hsu, P. S. S. (1981). *Chinese discovery of America*. Hong Kong: Southeast Asian Research Institute.
Hu, J. R., & Ma, D. F. (2000). Socialization and values among those from one-child families. *Youth Studies, 12*, 42–45.
Kirkland, R. (1986). The roots of Altruism in the Taoist tradition. *Journal of the American Academy of Religion, 54*, 59–77.
Kirkland, R. (2002). Self-fulfillment through selflessness: The moral teachings of the *Daode jing*. In M. Barnhart (Ed.), *Varieties of ethical reflection: New directions for ethics in a global context* (pp. 21–48). New York: Lexington Books/Rowman and Littlefield.
Krippner, S. (2002). Conflicting perspectives on Shamans and Shamanism: Points and counterpoints. *American Psychologist, 57*, 962–977.
Lee, K., & Ashton, M. (2004). Psychometric properties of the HEXACO personality inventory. *Multivariate Behavioral Research, 39*(2), 329–358.
Lee, K., Ashton, M., Morrison, D. L., Cordery, C., & Dunlop, P. D. (2008). Predicting integrity with the HEXACO personality model: Use of self- and observer reports. *Journal of Occupational and Organizational Psychology, 81*, 147–167.
Lee, K., Ashton, M., Pozzebon, J. A., Visser, B. A., Bourdage, J. S., & Ogunfowora, B. (2009). Similarity and assumed similarity in personality reports of well-acquainted persons. *Journal of Personality and Social Psychology, 96*(2), 460–472.
Lee, Y.-T. (1995). A comparison of politics and personality in China and in the U. S.: Testing a "kernel of truth" hypothesis. *The Journal of Contemporary China, 9*, 56–68.
Lee, Y.-T. (2000). What is missing on Chinese-Western dialectical reasoning? *American Psychologist, 55*, 1065–1067.
Lee, Y.-T. (2003). Daoistic humanism in ancient China: Broadening personality and counseling theories in the 21st century. *Journal of Humanistic Psychology, 43*(1), 64–85.
Lee, Y.-T. (2004). What can chairs learn from Daoistic/Taoistic Leadership? An Eastern perspective. *The Department Chair, 14*(4), 25–32.
Lee, Y.-T. (2011). Social psychology of stereotyping and human difference appreciation. In S. Chen (Ed.), *Diversity management: Theoretical perspectives, practical approaches, and academic leadership* (pp. 33–46). New York: Nova Science Publishers.
Lee, Y.-T., Bumgarner, J., Widner, R., & Luo, Z.-L. (2007). Psychological models of stereotyping and profiling in law enforcement: How to increase accuracy by using more non-racial cues. *Journal of Crime and Justice, 30*(1), 87–129.
Lee, Y.-T., Han, A. G., Bryron, T. K., & Fan, H. X. (2008). Daoist leadership: Theory and application. In C. C. Chen & Y.-T. Lee (Eds.), *Leadership and management in China: Philosophies, theories and practices* (pp. 83–107). New York: Cambridge University Press.
Lee, Y-T., & Haught, H. (2013). *Daoist leadership in psychology: From quantitative measures to practical implications*. Unpublished manuscript. The University of Toledo, Ohio.
Lee, Y.-T., & Jussim, L. (2010). Back in the real world. *American Psychologist, 65*(2), 130–131.
Lee, Y.-T., Jussim, L., & McCauley, C. (1995). *Stereotype accuracy: Toward appreciating group differences*. Washington, DC: The American Psychological Association.
Lee, Y.-T., Jussim, L., & McCauley, C. (2013). Stereotypes as categories of knowledge: Complexity, validity, usefulness, and essence in perceptions of group differences. *Advances in Psychological Science, 21*, 1–21.
Lee, Y.-T., & Li, H. (2011). Spiritual beliefs and ethnic relations in China: A cross-cultural and social psychological perspective. In Z. Q. Zhu (Ed.), *The People's Republic of China: Internal and external challenges* (pp. 239–258). Hackensack, NJ: World Scientific Publishing Co.

Lee, Y.-T., McCauley, C. R., & Draguns, J. (1999). *Personality and person perception across cultures*. Mahwah, NJ: Lawrence Erlbaum Associates, Publishers.

Lee, Y.-T., McCauley, C., Moghaddam, F., & Worchel, S. (2004). *The psychology of ethnic and cultural conflict*. Westport, CT: Praeger.

Lee, Y-T., Norasakkunkit, V., Li, L., Zhang, J. X., & Zhou, M-J. (2008). Daoist/Taoist altruism and wateristic personality: East and West. *World Cultures eJournal, 16*(2), 1–19. Retrieved from http://repositories.cdlib.org/wc/worldcultures/vol16/iss2/art2/.

Lee, Y.-T., & Ottati, V. (1993). Determinants of ingroup and outgroup perception of heterogeneity: An investigation of Chinese-American stereotypes. *Journal of Cross Cultural Psychology, 24*, 298–318.

Lee, Y.-T., & Seligman, M. E. P. (1997). Are Americans more optimistic than the Chinese? *Personality and Social Psychology Bulletin, 23*(1), 32–40.

Lee, Y.-T., Takaku, S., Ottati, V., & Yan, G. (2004). Perception and interpretation of terrorism, justice and conflict: Three cultures and two sides of one coin. In Y.-T. Lee, C. McCauley, F. Moghaddam, & S. Worchel (Eds.), *The psychology of ethnic and cultural conflict* (pp. 217–234). Westport, CT: Praeger.

Lee, Y.-T., & Wang, D.-Y. (2003). Aboriginal people in Taiwan, continent China and the Americas: Ethnic inquiry into common root and ancestral connection. In X. Li & Z. Pan (Eds.), *Taiwan in the twenty-first century* (pp. 63–82). Lanham, MD: University Press of America.

Lee, Y.-T., Yang, H.-G., & Wang, M. (2009). Daoist harmony as a Chinese philosophy and psychology. *Peace and Conflict Studies., 16*(1), 68–71.

Ma, H. K. (1992). The relation of altruistic orientation to human relationships and moral judgment in Chinese people. *International Journal of Psychology, 27*(6), 377–400.

Ma, H. K. (2003). The relationship of the family social environment, peer influences, and peer relationships to altruistic orientation in Chinese children. *Journal of Genetic Psychology, 164*(3), 267–274.

Ma, H. K., & Leung, M. C. (1991). Altruistic orientation in children: Construction and validation of the Child Altruism Inventory. *International Journal of Psychology, 26*(6), 745–759.

Meng, H. Y. (2005). 《寻找神秘的萨满世界》 *(xin zhao shen mi de saman shi jie) [Search for the mysterious Shamanic world]*. Beijing, China: Xiyuan Publishing House.

Moghaddam, F. (1998). *Social psychology: Exploring universals across cultures*. New York: W. H. Freeman.

Nagel, T. (1970). *The possibility of altruism*. Princeton, NJ: Princeton University Press.

Post, S. G. (2003). *Unlimited love: Altruism, compassion, and service*. Philadelphia, PA: Templeton Foundation Press.

Seyin, H. (1998). 《东北亚的萨满教》 *(Dongbei ya de saman jiao) [Shamanism in Northeastern Asia]*. Beijing, China: Chinese Social Science Academy Press.

Sima, Q. (1994/Han Dynasty or c. 150 B.C.). 《史记》 *(Shi Ji)* [*Records of the Grand Historian of China*]. Yin-Chuan, China: Ninxia People's Press.

Sorokin, P. A. (2002/1954). *The ways and power of love: Types, factors, and techniques of moral transformation*. Philadelphia. PA: Templeton, Foundation Press.

Tang, T. L.-P., Furnham, A., & Davis, G. M.-T. W. (2002). The meaning of money: The money ethic endorsement and work-related attitudes in Taiwan, the USA and the UK. *Journal of Managerial Psychology, 17*(7), 542–563.

Todd, E. (2003). *After the empire: The breakdown of the American order*. New York: Columbia University Press.

Triandis, H. C. (2001). Individualism-collectivism and personality. *Journal of Personality, 69*(6), 907–924.

Van Knippenberg, B., & van Knippenberg, D. (2005). Leader self-sacrifice and leadership effectiveness: The moderating role of leader prototypicality. *Journal of Applied Psychology, 90*(1), 25–37.

Watts, A. W. (1961). *Psychotherapy east and west*. New York: Pantheon Books.

Watts, A. W. (1975). *Tao: The watercourse way*. New York: Pantheon Books.

Wing, R. L. (1986). *The Tao of power: A translation of the Tao Te Ching by Lao Tzu*. Garden City, NY: A Dolphin Book/Doubleday and Co.

Xu, X. Z. (1991). 《山海经探》 *(Shan Hai Jing Tan) [The origin of the book Shan Hai Jing]*. Wuhan: Wuhan Publishing House.

Xu, P.-P., & Lu, A. Y. (2012). Rethinking of altruism moral education. *Social Science Academic Journal of the Science and Technology University of Central China, 26*(2), 37–42.

Yik, M. S. M., & Bond, M. H. (1993). Exploring the dimensions of Chinese person perception within indigenous and imported constructs: Creating a cultural balanced scale. *International Journal of Psychology, 28*(1), 75–95.

Yuan, K. (1988). 《中国神话史》 *(Zhongguo Shen Hua Shi) [Chinese History of Mythology]*. Shanghai: Shanghai Literature Publishing House.

Zhang, K., Lee, Y.-T., Liu, Y. F., & McCauley, C. R. (1999). Chinese-American differences: A Chinese view. In Y. T. Lee, C. R. McCauley, & J. Draguns (Eds.), *Personality and person perception across cultures*. Mahwah, NJ: Lawrence Erlbaum Associates.

Chapter 8
Altruism in Indian Religions: Embracing the Biosphere

Abhik Gupta

8.1 Introduction

Altruism, defined as unselfishness, an act of benevolence or welfare to others, or in a broader sense, cooperative behavior, still remains an inadequately understood trait in both animals and humans, despite the expositions of Hamilton (1964a, 1964b), Trivers (1971), Axelrod and Hamilton (1981), and others. The role of religion in nurturing and spreading altruism in human societies has been the subject of frequent and intense debate, often to the point of acrimony among psychologists, sociologists, evolutionary sociobiologists, brain and behavioral scientists, religious studies researchers, religious leaders, monks, priests, and the like.

Altruism becomes more enigmatic when it crosses the species barrier and is bestowed on nonhumans. Yet the occurrence of biophilia, which has been defined as the "innate tendency to focus on life and lifelike processes" (Wilson, 1984, as cited in Gullone, 2000) and the "innately emotional affiliation of human beings to other living organisms" (Wilson, 1993, p. 31), is common in humans and is on the rise in many societies. Wilson (1993, p. 31) has further said that "Innate means hereditary, and hence part of ultimate human nature." On the other hand, the lack of human altruism towards other species, even higher animals, has led to the extinction of species such as the dodo and the passenger pigeon (Cheke, 2006; Johnson, Clayton, Dumbacher, & Fleischer, 2010).

These extinctions, and the decimation even of magnificent creatures like the tiger, are poignant reminders of human ruthlessness towards other organisms. The Cartesian worldview of "man-nature dualism" that was popular for a long time in the Western societies, along with the interpretation of the biblical "dominion over nature" not as benign stewardship but as the birthright of humans to harvest plants,

A. Gupta (✉)
Department of Ecology & Environmental Sciences, Assam University,
Silchar, Assam 788011, India
e-mail: abhik.eco@gmail.com

D.A. Vakoch (ed.), *Altruism in Cross-Cultural Perspective*,
International and Cultural Psychology, DOI 10.1007/978-1-4614-6952-0_8,

kill animals, and harness all of the resources of nature without any inhibition and prudence, is often held (at least partly) responsible for this "human predicament" (Sessions, 1995).

On the contrary, many Eastern religions, including those which originate or are practiced in India, are characterized by the altruistic treatment of plants, animals, and even entire ecosystems and landscapes such as rivers, forests, and mountains, and recognize them as "kin" or hold them sacred and inviolable (Gupta & Guha, 2002; Singh, Singh, & Gupta, 2003). Such altruism could probably be called "biosphere altruism," to distinguish it from the altruism that is exhibited among members of the same species. Some "ecocentric" religions such as Jainism and Buddhism offered radical alternatives to more orthodox faiths that were or became over time lacking in empathy and compassion not only towards nonhuman plants and animals, but towards some fellow humans as well. The concept of *ahimsa* (nonviolence) to all living beings is an integral part of these religions.

Some other religions evolved among small groups of "ecosystem people" (Dasmann, 1988, p. 277) whose survival was intricately linked with the flourishing of the components of nature in their immediate vicinity, and whose cultural bonds with nature went as deep as, or perhaps deeper than, their economic bonds. Their worldviews, therefore, make no distinction among humans and the various animate and inanimate entities of nature. Their religious rituals and taboos, their folklores, and cosmogonies all reflect the perceived oneness of human and nonhuman worlds (Saraswati, 1993).

8.2 Altruism and the Concept of Value

Indian religions, which show altruism towards both humans and nonhumans, can be said to reflect the value ethics integral to their belief systems. As is well known, values are often classified into two basic types—extrinsic or instrumental value, and intrinsic or inherent value. Extrinsic value derives from the objective properties of something or somebody that is considered useful by the person who assigns the value. Thus, such value is entirely dependent on the "use value" of objects, plants, animals, or even human beings that is perceived by the value assigner. If plants, animals, or ecosystems (or other humans) are considered to be of value simply because they have some concrete material use for the value assigner, then such a value is merely extrinsic or instrumental. Consequently, the receivers of extrinsic value can cease to have any value if at any point of time they are no longer considered useful.

In contrast to extrinsic values, when we recognize intrinsic value in other living beings or nonliving entities, their value is independent of their utility to us. For example, most if not all of us recognize value in other human beings by virtue of their being conscious, intelligent creatures regardless of their utility or value to us (Martell, 1994). Such intrinsic value may be nurtured in association with some internal norm(s), which may be defined as a behavior pattern enforced by internal sanctions such as shame, guilt, sense of honor, empathy for others, and loss of self-esteem, and transmitted both vertically and horizontally. This is the opposite of external norms, which are driven by external sanctions such as rewards and punishments (Gintis, 2003).

External norms and extrinsic values, therefore, may be explained by the concept of reciprocal altruism—both direct and indirect—where the "give and take" between the altruist and the recipient of altruism is explicit. When relating the concept of values with human positions vis-à-vis plants, animals, and nature as a whole, we can say that recognizing only extrinsic or instrumental values in nonhumans is an anthropocentric approach, while even a tentative acceptance of intrinsic values in nature can be said to mark the beginning of ecocentric thoughts.

8.3 Altruism Towards Nonhumans in Indian Religions

8.3.1 Hinduism: From Sacrifice to Welfare

Hinduism has been described as the "dominant religious expression of the Indian subcontinent" with "a hoary mythology, an absence of recorded history (or founder), …… a pantheism that infuses divinity into the world around, and a tolerance of diverse paths to the ultimate God" (Swatos Jr., 2004). Thus, the roots of the altruistic attitude towards all living and even inanimate entities contained in the basic tenets of Hinduism lie in its pantheism and tolerance of diverse views. The earliest form of Hinduism was practiced by a group of people who called themselves "Arya," meaning noble. Their religious rites mainly revolved around large-scale animal sacrifice and fire ritual (*yajn*), which may not be regarded as eco- or even biocentric.

As the Aryans moved southeast from Northwest India, they came into a series of conflicts with the autochthonous, forest-dwelling aboriginals ("non-Aryans"), who practiced nature worship, maintained sacred groves, and can be regarded as having worldviews more in harmony with nature. In the Aryan-non-Aryan conflict, the latter were defeated and they either surrendered or fled to the more remote hills. Many were subjugated and subsequently assimilated into the Aryan fold. Around the same time, the sacrificial religion faced challenges from the aboriginal people, both Aryan and aboriginal ascetics (*yogis* and *munis*) and roving philosophers who were not only against wanton animal sacrifice and waste of resources but also disenchanted with the old religion that was becoming spiritually and philosophically hollow and relied on the pomp and grandeur of complex rituals. Around the same time, development of agriculture rendered large-scale sacrifice of cattle—which had more important use as milk and draft animals—untenable (Bhattacharji, 1984). Through the influence of all these factors, the sacrificial form of Hindu religion imbibed altruistic traits towards animals and other forms of life. Several plants and entire ecosystems such as mountains and rivers were also listed as sacred. The religion of the Upanishads recognized the presence of the "unchanging all-pervasive reality which is brahman" everywhere, in everything living or nonliving (Gosling, 2001). There was a general, omnipresent feeling of well-being for every entity—human and nonhuman, animate and inanimate—that found expression in a number of verses in these ancient texts.

One such verse from the *Isopanishada* says that "He who sees all beings in the self itself, and the self in all beings, feels no hatred by virtue of that (realization)" (Swami Gambhirananda, 1991, p. 13). Two other very well-known verses (*slokas*)[1] also prayed for the happiness and wellness of all and embraced everybody in the world as kin (Karan Singh as cited in Gosling, 2001). Such transcendence of Hinduism to a higher philosophical plane was perhaps reflected in the protection of "auspicious" birds that numbered around 17 species along with compassionate treatment of domestic animals recommended in *Kautilya Arthasastra* (KAS), an ancient Indian text written or compiled between third century BCE and second century CE. Its earlier stratum, which includes recommendations on issues of prudent utilization and conservation of resources and animal welfare, is believed to have been laid down in the third century BCE (Gupta & Ghosh, 2003). KAS advised cowherds to milk the cattle twice a day in the rainy season, autumn, and winter, but only once in the season of frost, spring, and summer, ostensibly because of the scarcity of sufficient amounts of nutritious fodder in the latter seasons. They were also required to treat calves and aged or sick cattle. It was suggested that for driving away cattle, people should use a length of rope or a thin switch so as not to hurt them. Similarly, forest animals, when found grazing in crop fields, were not to be killed or injured, but just driven away and the forest guards informed when necessary (Shamasastry, 1915). The caretakers of tame elephants were to be suitably punished if they kept the living quarters of elephants unclean, cheated on their daily ration, made them lie down on a rough or hard surface, hit them in tender and vulnerable parts of their body, allowed unauthorized persons to ride them, drove them at odd hours or on inhospitable ground, took them to water down steep riverbanks, or drove them to dense forests (Basak, 1964). Although the altruism shown here was linked to the high utility of elephants in warfare and transportation, the concern for the well-being of the elephants may be said to transcend beyond mere use value of the animals.

8.3.2 Influence of the Bhakti *Movement on Hindu Altruism*

The Upanishads imparted a spirit of philanthropy and altruism to Hindu religion and philosophy. Nevertheless, it is more than probable that the common people continued to practice the old rituals and that caste differences became increasingly ugly and exploitative. The *bhakti* movement that originated in South India, and later spread into the north in the medieval period, was a response to the orthodoxy and ritual-centric approaches, and instead preached the simple path of *bhakti*, which literally means

[1] One of the verses says: "Sarveh pi sukhinah santu/sarbe santu niramayah" (Let everybody be happy and free from disease); the other is: "vasudhaibo kutumbakam" [The whole world is (my) kin].

devotion. Along with an altruistic attitude towards all types of humans, compassion for all living beings was implicit in all the variants of the *bhakti* creed.

However, altruism to nonhumans found its most explicit expression in the neo-Vaishnavite movement of Srimanta Sankaradeva in fifteenth-century Assam. The cultural and religious scenario of Assam during that period was plagued by various degenerative practices such as animal and even human sacrifice, magical rites, and the like. The very integrity of the Assamese society was imperiled by this anarchy. At such a critical juncture of Assam history, the devotion-based preaching of Srimanta Sankaradeva (1449–1569) was highly effective in rescuing Assam from the quagmire of cultural and religious degeneration. Sankaradeva's religion did not recognize caste differences and treated all humans as equal. In fact, it went further to embrace all living organisms by finding soul or "god" or intrinsic value among nonhuman creatures as well. The result was a kind of religious unification of Assam that largely got rid of the degenerative practices that affected people (Das, 1978).

8.3.3 The Precept of Ahimsa *(Nonviolence) in Jainism*

Jainism evolved as a heterodox response to the sacrificial rituals of orthodox Brahmanism, which became increasingly complex and wasteful. One of the major precepts of Jainism is *ahimsa parmo dharma* (nonviolence/compassion is the supreme virtue/religion). Another important Jain dictum is *parasparopagraho jivanam*, which implies that living beings render service to one another or, in other words, depend on one another (Shah, 2009). The basis of these concepts extends to the Jaina definition of life, which includes both living beings and the elements themselves, an inclusiveness which in turn recognizes the inalienable relationships of humans with other living and nonliving components of nature.

These ideas are likely to be a more elaborate continuity of the Upanishadic image of a person as the reflection of the world in its entirety. The Jaina vision of the cosmos is that of a great human (female) body, the central realm of which is occupied by humans, animals, and other life-forms, with the heavens above and hell below. The Jaina classification of life-forms puts them in a graded array of five groups. Soil, water, fire, plants, and microorganisms are the first-order life, having only the sense of touch ("one sense"). On this base are laid down in succession the second-, third-, fourth-, and fifth-level beings with the last-named realm harboring birds, reptiles, mammals, and humans together (Umasvati, 1994, pp. 45–46, as cited in Chapple, 2001). Thus, the roots of Jaina altruism are grounded in this extended recognition of life that makes it possible to think beyond the human species. Such altruism automatically finds intrinsic value in all living organisms and even the inanimate components of nature. It is said that Lord Mahavira, the founder of the Jaina faith, exhorted his followers not to view the trees in terms of their economic or use value for making palaces, houses, furniture, plows, boats or carts, but as noble and magnificent life-forms extending their branches from a rounded trunk (*Akaranga Sutra* II.4.2.11–12, as cited in Chapple).

8.3.4 *Buddhism and the Concept of* Metta

Metta is a Pali word that denotes "loving-kindness," universal love, friendliness, kindness, goodwill, benevolence, amity, and the like. Metta has a negative connotation where it implies not doing harm to others; in its positive aspect, metta means doing good to all living beings. Stanzas 3, 4, and 5 in the Karaniya Metta Sutta, which is a universal hymn of love, say:

> May all be well and secure,/May all beings be happy/Whatever living creatures there be,/ Without exception, weak or strong,/Long, huge or middle-sized,/Or short, minute or bulky,/ Whether visible or invisible,/And those living far and near,/The born and those seeking birth,/may all beings be happy! (Buddharakkhita, 1989).

Metta is extended to all human beings be they friend or foe. Hence, by extending this concept to include other species, it embraces prey and predator, herbivore and carnivore, plant and animal alike. By including in its fold even the unborn creatures, metta may be visualized as spanning generations, thereby infusing a sense of sustainability in its precepts.

8.3.4.1 Emperor Asoka: Metta in State Policy

The Mauryas had established a vast empire across a large stretch of the Indian subcontinent during *ca.* 324–187 BCE. Theirs was a well-managed state with great emphasis on agricultural, forest, and livestock resources (Chakravarti, 1998; Rangarajan, 1987). Asoka (period of reign: 273/272–236/235 BCE), who had succeeded his father, Emperor Bindusara, was engaged in a devastating war to conquer Kalinga (more or less the present-day Orissa). There were heavy casualties on both the sides in the war. This is said to have moved Asoka so deeply that he eschewed violence and followed the teachings of the Buddha, which he was instrumental in spreading throughout Asia. He was probably the first ruler to take concrete steps to reduce animal slaughter and hunting and appears to have undertaken it in a phased manner. In the eleventh year of his reign, he promulgated the "Law of Piety," where he instructed his huntsmen and fishermen to stop or at least reduce hunting and fishing. He also drastically reduced the number of animals killed daily in the royal kitchen. In his seventeenth year on the throne, he started taking sterner measures, like expelling subjects who still indulged in the killing of animals and fishes, and in his twenty-seventh year, he brought several animals under state protection. Among other measures which deserve mention, he prohibited setting forests on fire so as to prevent death and injury to wildlife and hunting and fishing on certain days in different seasons, and encouraged the planting of trees, especially mango and banyan, digging wells for the benefit of men and beasts, and undertaking *dhamma yata* (religious tours) to make contact with his subjects and instruct them on the precept of nonviolence (Gupta & Ghosh, 2003; Mukherjee, 2000). The Rock and Pillar edicts of Asoka provide evidence for the fact that altruism towards nonhumans was not only confined to philosophical and religious teachings but was also incorporated into the policies of the Maurya state and disseminated among the people to remind them repeatedly to practice it in their daily lives.

Here is a brief description of the Rock and Pillar edicts (RE and PE, respectively) that contain altruistic proclamations towards nonhumans. In RE 1, Asoka declared that no living being was to be slaughtered or offered in sacrifice. While thousands of animals were earlier killed every day in the royal kitchen, their number was reduced to three, to be stopped altogether shortly. In RE 2, he made a provision, in his empire and in neighboring states, for medical treatment of humans and animals. In RE 3, he equated virtues such as respect for parents and generosity to friends, relatives, servants, and pious persons with not killing living beings. He reiterated the virtue of these practices in RE 4 and 11, praising them as an integral part of dhamma, which means piety, but may also connote ethical codes of conduct integrated into a political philosophy.

In his PE 2, Asoka asserted that he had given various things, including the gift of life, to two-footed and four-footed beings, to birds and aquatic animals. In PE 5, after 26 years on the throne, Asoka protected many terrestrial birds, such as parrots, doves and pigeons, and mynas; aquatic birds like geese and duck, other aquatic animals such as turtles, and several species of fish; mammals like porcupines, squirrels, deer, bulls, wild asses, and others. He said that the roadside plantations of banyan trees mentioned in PE 7 were for giving shade to both animals and humans, and he made watering places for animals (Dhammika, 1993). Among domestic animals, pregnant and weaning goats, ewes, and sows, as well as young ones less than 6 years old, were not to be killed. In addition, there were special days on the lunar calendar when slaughter of animals and fishing were totally prohibited (Gupta & Ghosh, 2003). Perhaps the most remarkable aspect of Asoka's compassion towards animals was his declaration that all four-footed creatures that were neither useful nor edible were not to be killed. This suggests that Asokan altruism was not motivated merely by the foresight to conserve resources that would be required in the near future, because animals not of any direct use to humans came under the umbrella of his compassion and protection.

8.3.5 "Biosphere Altruism" in Tribal Myths and Culture

Religious and cultural practices of the autochthonous communities of India comprised nature worship and offering protection to nature through the creation and maintenance of sacred groves, along with seasonal or other taboos on hunting, and other mechanisms of prudent resource use (Gadgil & Guha, 1992; Gadgil & Vartak, 1994; Gadgil, 1995). Maintenance of sacred groves, which are community-protected forests found throughout India, may be considered as examples of altruism towards plants, animals, and nature in general. In many sacred groves, total protection is accorded to all forms of life, and even removal of dead and decomposing wood is not allowed in many. Thus, sacred groves as an ancient institution can be said to have promoted an altruistic treatment of nature since long.

Ancestor worship is also common in these societies. Some groups do not harvest certain plants and abstain from killing certain animals, either totally or at least in certain seasons or life stages. Many communities exhibit totemism, where they do not harm the totem plant or animal, and undertake specific rites and rituals to propitiate them, which they consider to be kin or ancestors. Their myths and folklore also reflect this unity of humans and nature, devoid of any distinction among god, humans, plants, animals, and even inanimate natural entities. The creatures themselves play an important role in the creation and construction of the world. Man's origin can be traced to plants, animals, and inanimate objects, and there is kinship between man and these entities. These myths also deny the uniqueness of man in the possession of knowledge, as nonhumans frequently hand over vital knowledge to them (Saraswati, 1993).

A few examples from the tribal myths and folklores from Arunachal Pradesh, Northeast India, may illustrate these points further. Sacrifice is the basis of many tribal myths of creation found in Arunachal Pradesh. The Apatani tribe believes that *kujum-chantu* (mother earth) died voluntarily to lay the basis of creation, the different parts of her body forming the celestial bodies and other elements. Thus, a sense of "primordial altruism" ran deep in the tribal mind. The role of love in creation is also emphasized in a Hrusso Aka myth, which says that when the sky made love to the earth, all the trees and other living creatures came into being.

Several myths also describe the important role of nonhumans in creation and construction. The Hill Miris of Arunachal Pradesh believe that in the beginning, there was water everywhere, with only a gigantic tree raising its head above. A worm gnawed at its wood and the dust fell all around to form the earth. Finally the tree fell to the ground, and the bark on its lower side became the skin of the earth, while that on the upper side became the sky, and the trunk formed the rocks, and the branches became mountains. A Gallong myth says that the prawn collected a great pile of rubbish on the surface and the crab dug a hole to drain the water, and thus the land emerged (Elwin, 1958). Similarly, a story of creation prevalent among the Birhors of Bihar says that in the beginning when there was only water, and when God asked the tortoise to fetch some clay from the bottom of the ocean, it failed. The crab too failed, and finally it was the leech that dived into the ocean, had its fill of clay, and then came up and egested all the soil from which god created the land mass (Adhikary, 1995).

A number of myths also trace the nonhuman ancestry and kinship of man, such as from frogs (Dhammai Miji myth), or from a flower (Khampti myth). The great primeval spirits had three children, one a human, another a rock, and yet another a gourd. When the rock-child broke open the gourd, the first humans emerged (Singpho tale). In another story (Taraon Mishmi), the first men and women came out of the tusk of an elephant. There are many stories of the marriage and sexual union of humans with gods, spirits, real animals (and not humans in animal disguise) like snakes, monkeys, tigers, etc., and even leaves, trees, and fire. In these myths, man is not unique in the possession of knowledge, which more than often came to him from animals. Thus, the birds (Hill Miri myth) or flowers and bees (Bugun myth) taught the first man and woman the art of reproduction, the spiders taught a girl how to weave (Singpho), and the rat taught the technique of cultivation (Saraswati, 1993). In another story (Idu Mishmi), the sparrow taught cultivation to humans (Elwin, 1958). The Liangmai Nagas of Manipur tell a story in which the queen of the rats retrieved the bags of paddy that the humans in their carelessness had allowed to float into the sea. Because the rats had agreed to recover the bags, they are allowed to share the paddy produced by the humans (Miri, 1995).

8.4 Evolutionary Perspective

Understanding religion has probably been one of the greatest intellectual challenges that has continued for ages and yet has remained inconclusive and hotly debated. This chapter poses a relatively simple question, which nevertheless is embedded in the complex matrix of the origin and evolution of religion and its nature and implications as well. That question pertains to the purpose and significance of including nonhuman living and nonliving entities in the domain of human altruism in the various "nature religions," as well as in Jainism, Buddhism, and others, such as the neo-Vaishnavite creed of Srimanta Sankaradeva in Assam. The two contrasting views on religion comprise the adaptationist and non-adaptationist or by-product theories. The former regards religion as an adaptation or of direct adaptive value, while the latter maintains that religious concepts have piggybacked on other cognitive adaptations, especially those for agency detection. Thus, it is a by-product of adaptive cognitive structures (Sanderson, 2008). Sosis (2003) and Sosis and Alcorta (2003) argued that religious rituals could be regarded as costly and hence "hard to fake" signals that demonstrate the commitment of the signaler to the group. The costly signaling theory has since attempted to provide a plausible explanation for religious rituals. It may, therefore, be worthwhile to analyze the problem of the extension of religious altruism to nonhumans as observed in several Indian religions in the light of these theories, keeping in background the historical context of their appearance.

If we look at the emergence of Jainism, which espouses the principle of nonviolence to both humans and nonhumans, we can observe that it emerged at a critical juncture in Indian religious history when the old sacrificial brahmanical religion was facing a crisis of credibility. Its rituals, which had themselves evolved as costly signals, had become complex and wasteful, and resultantly too costly when weighed against the payoffs. Nevertheless, the Upanishadic reforms, which emphasized self-realization at a higher philosophical plane, were trying to salvage the situation as best as possible. Therefore, a new faith like Jainism needed honest and costly signals to convince people of its inclusiveness and compassion, both of which were served by including plants, animals, and even the natural elements within its sphere of inclusion and protection. Such signals must have attracted a sizeable number of adherents, especially from the *Vaisya* caste, made up of tradespeople, who cherished the message of peace, compassion, abstinence from violence, and the emphasis on mutual dependence (*parasparopagraho jivanam*).

A similar situation confronted Srimanta Sankardeva in fifteenth-century Assam, which was passing through a harrowing period of cultural and religious degeneration and anarchy. At the same time, it had a high ethnic diversity that somehow had to be kept together in a spirit of coexistence and cooperation under the umbrella of a common religion acceptable to all. A highly inclusive message had to be sent to the people to convince them of the compassion and openness of the new creed that did not recognize caste differences. One of Sankardeva's verses therefore vouches for the godliness of the soul of not only the lowliest of the lowly human but that of animals like dog and donkey as well. It was a signal that deliberately incorporated an overemphasis to appear honest and convincing, especially as the dog is traditionally considered "unclean" in the Hindu society, while the donkey represents naiveté.

The incorporation of altruism towards animals into state policies by Asoka could also be viewed as a costly signal. After the Kalinga war, Asoka must have reasoned that a nonviolent "dove" policy had a chance to succeed only if he could send a clear message to the neighboring rulers that his commitment to peace and nonviolence was genuine and his formidable war machine posed no threat to them. The prominent public display of his Rock and Pillar edicts was integral to this signaling, and stopping or at least drastically reducing animal slaughter and even the more innocuous yet more rampant practice of fishing made the signal costlier, requiring many sincere efforts for effective implementation. With the passage of time, he adopted sterner measures in order to further prove his honest intent. Based on the above examples, it may be hard to label altruism towards nature as a by-product of something else that was an adaptation. Rather it appears to be of adaptive value by serving as costly signals to render strategies successful.

References

Adhikary, A. K. (1995). The Birhor universe. In B. Saraswati (Ed.), *Primal elements: The oral traditions*. New Delhi: Indira Gandhi National Centre for the Arts.

Axelrod, R., & Hamilton, W. D. (1981). The evolution of cooperation. *Science, 211*, 1390–1396.

Basak, R. G. (1964). *Koutilya Arthashastra (in Bengali)*. Kolkata: General Printers and Publishers.

Bhattacharji, S. (1984). *Literature of the Vedic Age Vol. 1: Samhitas*. Kolkata: K.P. Bagchi and Co.

Buddharakkhita, A. (1989). *Metta – The philosophy and practice of universal love*. DharmaNet Edition 1995 (The Wheel Publication No. 365/366). Kandy, Sri Lanka: Buddhist Publication Society; Berkeley, CA: DharmaNet International. http://enlight.lib.ntu.edu.tw/FULLTEXT/JR-AN/an140977.pdf.

Chakravarti, R. (1998). The creation and expansion of settlements and management of hydraulic resources in ancient India. In R. H. Grove, V. Damodaran, & S. Sangwan (Eds.), *Nature and the orient* (pp. 87–105). New Delhi: Oxford University Press.

Chapple, C. K. (2001). The living cosmos of Jainism: A traditional science grounded in environmental ethics. *Daedalus, 130*, 207–224.

Cheke, A. S. (2006). Establishing extinction dates – the curious case of the Dodo *Raphus cucullatus* and the Red Hen *Aphanapteryx bonasia*. *Ibis, 148*, 155–158.

Das, K.C. (1978). An approach to Assam vaishnavism in the light of the Upanishads. In B.P. Chaliha (Ed.), *Sankaradeva: Studies in culture, 2nd Edition, 1998*. Assam: Srimanta Sankaradeva Sangha. http://www.atributetosankaradeva.org/approach.pdf.

Dasmann, R. F. (1988). Toward a biosphere consciousness. In D. Worster (Ed.), *The ends of the earth: Perspectives on modern environmental history* (pp. 277–288). Cambridge: Cambridge University Press.

Dhammika, V.S. (1993). *The Edicts of King Ashoka* (The Wheel Publication No. 386/387). Kandy, Sri Lanka: Buddhist Publication Society; DharmaNet Edition 1994. Berkeley, CA: DharmaNet International. http://www.cs.colostate.edu/~malaiya/ashoka.html#PREFACE.

Elwin, V. (1958). *Myths of the north east frontier of India*. Shillong: North East Frontier Agency.

Gadgil, M. (1995). Prudence and profligacy: A human ecological perspective. In T. M. Swanson (Ed.), *The economics and ecology of biodiversity decline: The forces driving global change* (pp. 99–110). Cambridge: Cambridge University Press.

Gadgil, M., & Guha, R. (1992). *This fissured land*. Delhi: Oxford University Press.

Gadgil, M., & Vartak, V. D. (1994). The sacred uses of nature. In R. Guha (Ed.), *Social ecology* (pp. 82–89). Delhi: Oxford University Press.

Gintis, H. (2003). The hitchhiker's guide to altruism: Gene-culture coevolution, and the internalization of norms. *Journal of Theoretical Biology, 220*, 407–418.

Gosling, D. L. (2001). *Religion and ecology in India and Southeast Asia*. London: Routledge.

Gullone, E. (2000). The biophilia hypothesis and life in the 21st century: Increasing mental health or increasing pathology? *Journal of Happiness Studies, 1*, 293–321.

Gupta, A., & Ghosh, S. (2003). Wildlife conservation in the Maurya period. *Journal of Ancient Indian History (2000–2002) (Kalyan Kumar Dasgupta Memorial Volume), 21*, 142–152.

Gupta, A., & Guha, K. (2002). Tradition and conservation in Northeastern India: An ethical analysis. *Eubios Journal of Asian and International Bioethics, 12*, 15–18.

Hamilton, W. D. (1964a). The genetical evolution of social behaviour. *International Journal of Theoretical Biology, 7*, 1–16.

Hamilton, W. D. (1964b). The genetical evolution of social behaviour. *International Journal of Theoretical Biology, 7*, 17–52.

Johnson, K. P., Clayton, D. H., Dumbacher, J. P., & Fleischer, R. C. (2010). The flight of the passenger pigeon: Phylogenetics and biogeographic history of an extinct species. *Molecular Phylogenetics and Evolution, 57*, 455–458.

Martell, L. (1994). *Ecology and society: An introduction*. Cambridge: Polity Press.

Miri, S. (1995). The order of nature in Liangmais' myth. In B. Saraswati (Ed.), *Primal elements: The oral traditions*. New Delhi: Indira Gandhi National Centre for the Arts.

Mukherjee, B. N. (2000). *The character of the Maurya empire*. Calcutta: Pilgrim.

Rangarajan, L. N. (1987). *Kautilya – the Arthashastra*. New Delhi: Penguin.

Sanderson, S. K. (2008). Adaptation, evolution, and religion. *Religion, 38*, 141–156.

Saraswati, B. (1993). The implicit philosophy and worldview of Indian tribes. In G. Fløistad (Ed.), *Contemporary philosophy: A new survey* (Asian philosophy, Vol. 7, pp. 121–136). Dordrecht: Kluwer Academic Publishers.

Sessions, G. (1995). Ecocentrism and the anthropocentric detour. In G. Sessions (Ed.), *Deep ecology for the 21st century* (pp. 156–183). Boston: Shambhala.

Shah, P. (Ed.) (2009). *Jainism – Religion of compassion and ecology (Seventh Edition – Expanded)*. Jaina Education Series (JES 901), Jaina Education Committee, Federation of Jain Associations in North America, Jain Study Center of North Carolina (Raleigh). www.jaina.org.

Shamasastry, R. (1915). *Kautilya's Arthasastra, Seventh Edition, 1961*. Mysore: Mysore Printing and Publishing House.

Singh, L. J., Singh, N. B., & Gupta, A. (2003). Environmental ethics in the culture of Meeteis from North East India. In S. Y. Song, Y. M. Koo, & D. R. J. Macer (Eds.), *Bioethics in Asia in the 21st century* (pp. 320–326). Tsukuba: Eubios Ethics Institute.

Sosis, R. (2003). Why aren't we all Hutterites? *Human Nature, 14*, 91–127.

Sosis, R., & Alcorta, C. (2003). Signaling, solidarity, and the sacred: The evolution of religious behavior. *Evolutionary Anthropology, 12*, 264–274.

Swami Gambhirananda (Tr.). (1991). *Eight upanisads: With the commentary by Sankaracharya* (Vol. 1). Calcutta: Advaita Ashrama.

Swatos, W.H. Jr. (Ed.) (2004). *Encyclopedia of religion and society*. Hartford Institute for Religion Research, Hartford Seminary. Retrieved July 8, 2012, from http://tdissmeyer.com/10%20-%20Hinduism/Hinduism%20by%20Ency%20of%20Religion.pdf.

Trivers, R. (1971). The evolution of reciprocal altruism. *The Quarterly Review of Biology, 46*, 35–57.

Wilson, E. O. (1993). Biophilia and the conservation ethic. In S. R. Kellert & E. O. Wilson (Eds.), *The biophilia hypothesis* (pp. 31–41). Washington, DC: Island Press.

Chapter 9
Altruism, Renouncing the Renunciant: A Discussion on Desire and Selflessness

Sangeetha Menon

9.1 Introduction

The origin of the term altruism is more Western than Eastern in its import and usage. Comte, the father of sociology, adopted it to describe dispositions, tendencies, and actions that have the good of others as their object. According to Comte, the goal is not to subordinate egoism to altruism, but to develop egoism to its proper proportions, in the belief that the higher and fuller a personality is, the more it has to contribute to the happiness of humankind (Iverach, 1994). The mainstream explanations for altruism have been biological and support rules of adaptation. Using the selfish meme approach to explain all of the past, present, and future of human mind and creativity is a reductive epistemological strategy to biologically explain self-expressions, such as altruism, that have spiritual roots (Menon, 2002). Behavioral sciences that inspire sociobiology theorize the nonexistence of a nonphysical self with arguments that are based on mimetic information and our evolutionary stance. Self is nothing but a bunch of memes (Dawkins, 1976) and is constituted by the genetic processes behind the transmission of cultural traits. The dualistic division of the human self into acts such as "selfishness" and "altruism" is restrictive and offers limited frameworks to explain the complexity of subtler human expressions.

Arguments that favor selfishness as a natural, biological trait that supports species survival are meaningful when limited to body-oriented experiences and personal identity that are defined by constraints such as the physical body, limited resources, distrust of members of the same species, and fear of powerful predators. The nature of selfishness is different when it comes to a species that is defined primarily by its culture and less by biology. A marked feature of the human species is the complexity in self-perception and perception of the other. "Me" and "you" is a central divide

S. Menon (✉)
National Institute of Advanced Studies, Indian Institute of Science Campus, 560012 Bangalore, India
e-mail: smenon@nias.iisc.ernet.in; prajnanata@gmail.com

D.A. Vakoch (ed.), *Altruism in Cross-Cultural Perspective*,
International and Cultural Psychology, DOI 10.1007/978-1-4614-6952-0_9,

that is overtly and covertly present in human living and engagement. This divide is the motivator for desire, action, and the exercise of choice making. Sociobiological discussions of altruism give exclusive attention to altruism as an *act* with evolutionary or social benefits. It is not considered that altruism is a phenomenon exhibited by a living *self* with mind and emotions.

A good place to begin a dialogue on "selfishness versus altruism" is to ask "who is selfish for whom?" and "who is altruistic for whom?" If the responses are not self-revealing, then the next tier of questions is "why selfish?" and "why altruistic?" And the subsequent tier of questions posits a fundamental context for a dialogue focused on identity and embodiment. A resolution of the seemingly opposing duals of "take in" and "give up" is possible only if we elevate the discussion from an exclusive sociobiological space to the space of a deeper, core self.

In the context of Indian wisdom traditions, the concerns of altruism are best understood against a background of the concepts self, desire, and well-being. The representations of self, identity, and the ability to give up and share are best explained by the ontological considerations of the self. Altruistic and selfish acts are driven by the self-concepts and the sense of security that emerges from the (inclusive or exclusive) space of the self. The Vedantic idea of self as pure consciousness posits to possess a space that is unlimited and all-inclusive. From the *atman* space, no acts are directed because of the binary of "selfishness and altruism" but are gauged toward the well-being of all. There is neither giving up for the sake of another nor possessing what belongs to another. All acts are intended for mutual nourishment. There is no giver and taker, from the standpoint of a self that is established in true identity, according to Vedanta.

In Indian traditions, and Vedanta in particular, the construal of self suggests that the concept of *spiritual altruism* is oriented toward the greater and common good of the body, mind, and the spirit. The nature of selflessness in spiritual altruism emerges from the selflessness as a *state of being*. It is directly connected to the transformation of consciousness, influencing compassion, empathy, and social good.

The question to be debated is whether altruism and selfish behavior are better understood if we make a deliberate shift of focus from the act, and behavior, as articulated in preservationist, hedonistic theories evidenced in the sociobiological literature, to formation of self-identities, group identities, and the process of self-transformation (Menon, 2007). The concept of spiritual altruism is best understood with the help of some examples from Indian wisdom traditions that inspire us to go beyond the simple binary of selfishness and altruism.

9.2 A Story from the *Mahabharata*

How do we respond when confronted with grave risks and minimal options? What is considered a priority when life is at stake? How much self-giving is motivated by the urge to rescue kith and kin? The impact of choice and decision-making in defining risk is best illustrated by a story from the *Mahabharata*, which narrates the

Pandava brothers' rendezvous with Yaksha. The story goes thus (narration is adapted from Subramaniam, 1990):

> Once, when the Pandavas were living in Dwaitavana—it was during their final stay of a few months—a brahmin came to Yudhishthira with an appeal. He said that a deer had entered his hut and carried away the sticks used for making fire, the Arani. The Pandavas left at once and went in pursuit of the deer. They followed it very far but suddenly it disappeared from their sight. Depressed in mind and fatigued by thirst and hunger, they sat down under the shade of a huge tree. They were very unhappy and all the brothers except Yudhishthira started lamenting about the fate and the unending number of woes that follow them.
>
> Yudhishthira smiled at them all and said, "This is not the time to look back and think of the might-have-beens. Our immediate worry is this: How are we to quench this dreadful thirst that has been troubling us since some time? Nakula, get up on the tree and look around. See if you can find any spot of water in the neighborhood. We are all almost dying with thirst." Nakula did as he was told. He said, "I can see a lake just nearby." They were all so happy to hear it. Yudhishthira said, "My child, go at once and bring water for all of us." Nakula hurried to the lake.
>
> He reached the lake. The water looked so cool and inviting. He went near it to drink it. Suddenly he heard a voice from nowhere. It said, "You must not drink the water of this lake, not before you have answered certain questions of mine. You can neither drink this water nor can you carry it unless you answer my questions." Nakula did not pay any heed to this voice coming from nowhere. He was very thirsty. He rushed to the brink of the lake and drank the cold water eagerly. Immediately Nakula fell down dead. The others waited for him for a long time. But Nakula did not return.
>
> Yudhishthira sent Sahadeva to go in search of his brother. Sahadeva reached the lake. He saw the dead form of his brother on the ground. He was shocked at the sight. But his thirst was so great that he rushed towards the water as Nakula had done. The same voice was heard with the same warning. But Sahadeva was like Nakula. He disregarded the warning and drank the water and suffered the same fate as his brother. Yudhishthira next sent Arjuna and then Bheema. Not one of them came back. Yudhishthira waited for a long time and yet they did not come back. Intrigued by this, and with misgivings of the mind, Yudhishthira walked towards the lake. He reached it soon. He stopped in his tracks, horrified by the sight that met his eyes. He saw all his brothers there, dead.
>
> Yudhishthira was almost mad with grief. His roving eyes fell on the cool water and his thirst came back. His throat was parched and dry with unshed tears. He walked to the brink of the lake and was about to drink the water, when he was arrested by the unearthly voice. He was told that he should not drink until some questions were answered. Yudhishthira paused in the act of drinking. He looked around to locate the source of the voice. The voice said, "I saw your brothers come here one by one. I told them not to drink. They would not listen to me. They drank and died. I am the Yaksha who owns this lake."

The story goes on to explain that Yudhisthira answers all of the 60 odd questions posed by Yaksha, and to Yaksha's happiness he is not only granted water from the lake but also the lives of the four brothers. In this event, Yudhisthira and his brothers had to face two risks: the risk of having to die with no water to drink and the risk of having to die by giving wrong or no answers to Yaksha. There was the risk of death in both options of choices and decisions. Yudhisthira in this story places his life at risk to rescue the lives of others. And that is why it was prudent for him to strategize how he should be preserving his own life.

The brothers were already fatigued by hunger and afflicted by depression. Since their immediate worry was thirst, Yudhisthira proded his brothers to find a water source. Nakula and the other brothers found a lake and confronted the owner of the lake—the Yaksha—but did not heed his questions. Because of the blinding desire for at least a few drops of water, the surroundings did not matter much to them. They were unable to focus on what was happening in their environment. The water of the lake was so cool and inviting that even death did not become a concern. What obsessed them deeply was a haste to somehow quench their thirst. Hence, they either did not hear the voice of Yaksha or ignored him and his forewarning of death. At that point, quenching thirst was the primary desire. Even the sights of the dead bodies of the brothers did not stop Bhima, Arjuna, and Sahadeva from stepping to the lake and drinking water.

Only Yudhisthira heard the voice of Yaksha, in spite of his desire to quench his thirst, and had the patience as well as courage to answer to Yaksha's questions. What we may conclude from the story is that when pushed into a corner, we are forced to take a risk, even if it is at the cost of a trade-off like death. All acts are directed toward one's well-being.

9.3 Maitreyi's Dilemma and the Three *Da*

Brhadaranyaka Upanishad presents a dialogue between a husband and wife. Yajnavalkya is a mendicant and sage, and Maitreyi, his wife, an earnest philosopher. Yajnavalkya decides to pursue a secluded life and leave his family. Maitreyi, unsure about her choice, makes an effort to stop him by asking what is worth having. This dialogue raises two issues: when does the "other" give you up and what makes the "other" desirable. The discussion's center point is *atmanastu kamaya sarvam priyam bhavati*—that is, everything else becomes endearing because of the endearment to Self.

Maitreyi's dilemma was what use are material benefits if they cannot ensure immortality? In response to her query, Yajnavalkya gives a series of instances for "cherishability." The pertinent question implied in this dialogue is whether we can discretely possess and give up anything. Is that which is possessed and given up always neatly demarcated? The fringes of "me" and the "other" are so delicately separated that the distinction itself is created by human interventions like culture, social, and individual demands. The "other" that which is given up, according to Yajnavalkya, is never given up in the truest sense since the "other" is not a real entity. The most valuable possession is the Self which includes everything, even the "giver" that makes the demarcation between the other and oneself, and that which is possessed and that which is given up. Any act or possession becomes meaningful because it adheres to a self. The Vedantic interpretation of this discussion is that the source of contentment that is invoked by an object (or person, or relation) inheres in the Self. According to this framework of Self, there are no generalist altruistic or selfish acts. It is one's social and individual responsibilities and adherence to one's set of values that design an act to be altruistic or selfish.

The *Mahabharata* and the Indian wisdom traditions introduce the concept of *dharma* to point toward the primacy of individual uniqueness as a criterion that is prior to a labeling based on generalist standards and perceptions. An altruistic act performed by one could also be interpreted as selfish if seen out of context of the *dharma* of a particular individual. Altruism is to be understood from the context of one's *dharma* (Davis, 2005, p. 164).

The contextual significance of act performed and objects given up is illustrated by another story in the Brhadaranyaka Upanishad. The representatives of three classes of beings—humans, demons, and demigods—go to the god of creation, Brahma, to learn about best practices in life. Brahma's monosyllabic response to all of them was *da*. The syllable *da* invoked three different meanings in the minds of the humans and demigods. Humans understood *da* as *dana*, charity; demons understood *da* as *daya*, compassion; and demigods understood *da* as *dama*, control of mind.

While we discuss whether a particular act is altruistic or not, it is to be considered from the context of one's ability to give and share, to be compassionate, and to restrain one's desires. Altruistic expressions are borne from the continuum of the person whose identity is defined by the processes and range of the three abilities—sharing, compassion, and restraint of desires (desire being the innermost component of mind). In a sense, what we consider as an expression of altruism is perhaps not so, but is only a sensitivity to detach from what belongs to another. Such an idea is expressed in the *Isavasya Upanishad.* The opening verse of this Upanishad exhorts to discriminate between what belongs to oneself, and what belongs to another, and to consume only what belongs to oneself, and not to have greed for another's legitimate possession.

9.4 Desire and Its Mysterious Functions

The *Bhagavad Gita* and the *Kama Gita*, in the *Mahabharata,* place the psychology of desire as the forerunner to understand the philosophy of any act of renunciation. While the *Bhagavad Gita* details the dynamics of desire with the help of the agent of actions, the *Kama Gita* mocks human efforts to defeat the indefatigable nature of desire and its existence.

We usually understand "desire" as a propensity of the mind to attain a desirable object or to move away from something which is undesirable. The *Kama Gita* satirically presents another form of desire. Desire is portrayed as the intentor behind the intent of "desiring." Even when we think that we have transcended a particular desire, desire persists in another form which is the agent behind such a thought. Far from being a propensity, an act, and a behavior, the dynamics of desire is deep rooted in agenthood—one who desires and one who gives up a desire.

The idea is that true giving up happens not by giving up an object, or a person, or any physical attribute, but the agent who intends the process of giving up. Such a focus on agenthood is not be misinterpreted as an unethical dismissal of responsibility. Giving up the giver is not even a mental or physical act, but a spiritual act of extending one's self for maximal inclusion, of all duals. The object of renunciation is not the "other" but is the limited self-identity—this idea is at the core of the philosophy of action in the *Bhagavad Gita*.

The *Bhagavad Gita* discusses the concepts of desire and "giving up" in the background of a discussion on agency. A radical view presented in the *Gita* is that renunciation is a state of being and not necessarily an act. The habitual nature of mind is to initiate an action motivated by a desire and later to get perturbed by the outcomes. The root of desire is the attachment of mind to objects in the form of expectation for contentment. Attachment to objects marks a chain of psychological mishaps, according to the *Gita*.

The *Bhagavad Gita* explains the nature of desire and its varied expressions based on the theory of *guna* or individual propensities. The *guna* theory of *Bhagavad Gita* is a typology built on attitudes and dispositions. The *Bhagavad Gita* advocates a vocation in tune with one's *prakrti,* which is called *svadharma*. *Prakrti* is a dynamic of the three *gunas—sattva*, *rajas*, and *tamas*. The *prakrti* of an individual is determined by the predominant *guna* in his mental makeup. Accordingly, he is inclined to different pursuits/vocations. His ego, knowledge, work, inclinations, goals, happiness experience are all determined by this *guna* mix. In the area of mind and work, the *Bhagavad Gita* is deterministic. There is no way a *tamasic* mind can become *rajasic* and then *sattvic* as far as choice of vocation is concerned. That is why different disciplines are prescribed for *brahmana*, *kshatriya*, *vaisya*, and *sudra*. One's vocation is predetermined, depending on the *prakrti*, which is a carry forward from the past. But this has nothing to do with one's birth in a particular social stratum, but is purely by the inborn *guna* and the consequent *karma* (Bodhananda, 2005).

The *Gita* emphasizes that a person of deluded intelligence, who mistakes the Self as the agent and the enjoyer, understands the truth neither of the Self nor of the action: "He who is untrained in understanding, looks on the pure Self as the agent, that man of perverted intelligence sees not" (*Bhagavad Gita*: 18.16). He becomes bound by "the threefold fruit of action—evil, good and mixed" (18.12). But he whose understanding is untainted and attitude is non-egoistic, though he works, is not bound by his work. "He who is free from egoistic notion, whose mind is not tainted ...He is not bound" (*Bhagavad Gita*: 18.17). For him, work becomes a medium for creative expression.

9.5 Is "Altruism" a Cover-Up for Escapism?

The dialogue of the *Bhagavad Gita* begins with a breath-taking description of the might and prowess of two armies in the battlefield. As narrated by Sanjaya to Dhritarashtra, there is a detailed description of the men, weaponry, and relative strengths of the opposing armies put in the words of Duryodana to Drona. Having seen the army, Arjuna asks Bhagavan Krishna to take his chariot to the middle of the two armies so that he can make a better visual assessment. The view of the armies has a huge physical and psychological impact on Arjuna because he sees in both armies fathers, grandfathers, teachers, maternal uncles, brothers, sons, grandsons and comrades, fathers-in-law, and friends. The massive assemblage of chariots and his kith and kin as fighters on both sides in the vast arid field of Kurukshetra, his precarious placement between the armies, the uncertain outcome of the battle—all these might have brought down Arjuna's sense of identity from the esteemed heights of a warrior to the depths of a lesser mortal. What follows are disconnected words from a grief-stricken Arjuna, hopelessly hoping to avoid war and ready to quit the battlefield and lead a life of renunciation. Arjuna sees the meaninglessness of all that which he possessed and valued until then. Overcome by self-destructive thoughts, Arjuna talks about giving away all his wealth, land, and precious belongings and killing himself as a good option. Finally, we see an Arjuna ready to flee the scene.

In the beginning of the *Gita*, we see Arjuna ready to give up everything—the battlefield, wealth, victory, and so forth—and quit the need to act in a way he chose earlier. Was Arjuna being altruistic, ready to give up everything for his kith and kin and even to his foes? The rest of the dialogue in the *Gita* delves into the attitude which Arjuna thought he very nobly possessed at that point of time—willingness to give up victory and fame—for the same of others. Krishna in the course of the dialogue shows to Arjuna, with fine philosophical arguments that drive deep into his mind, how flimsy, deceptive, and hypocritical his notion of altruism and selflessness are.

Every act and attitude of altruism carries an apparent stamp of giving up something precious. But the analysis of the person's identity, value system, and the experiential crises people undergo reveals the intricacies and truth of such acts and attitudes. Many times we are ready to give up that which we either are tired of, or from which we wish to escape.

9.6 Give Up the Giver

Renunciation is not the physical giving up of action or the outcomes of action, according to the *Gita*, but is the essential attitude for the performance of an action. It is the subtle awareness that springs from the discrimination of the Self and the not-Self. Renunciation is the quality of mind that promotes true action. S. Radhakrishnan, the noted compiler of Indian philosophies, writes:

> Inertia is not freedom. Again, the binding quality of an action does not lie in its mere performance but in the motive or desire that prompts it. Renunciation refers, not to the act itself but to the frame of mind behind the act. Renunciation means absence of desire. So long as action is based on false premises, it binds the individual soul. If our life is based on ignorance however altruistic our conduct may be, it will be binding. The *Gita* advocates detachment from desires and not cessation from work. (1977, pp. 67–68)

Renunciation is the discarding of the false notion that "I am the doer" and "I am the enjoyer." It is not the action that is to be renounced, but the false notion of egoistic agency polluting its quality. It is not even giving up a particular behavior or giving up the urge to protect the self. Self-renunciation is not self-loss as Welsh and Knabb argue:

> … that concern for the preservation of the self is a pathogenic inheritance common to all humankind and an integral part of development. Specifically, selfishness is acting consciously or unconsciously with the intent of protecting oneself from a perceived external or internal threat. Threats may be real or perceived and can easily be witnessed by observers of newborns who are afraid or angered by light, sounds, and other unwanted sensations.

> The infant will react with wincing, crying, or some other behavior that is intended to protect the fragile newborn. There is an innate sense that the world is dangerous and that the vulnerable self must be protected. It is intriguing to note that nearly all psychological difficulties will result in an attempt to save oneself. (Welsh & Knabb, 2009, p. 407)

Renunciation is not escaping from the fear of self-loss, or an act of giving up any particular act, but is the detached and equanimous response to the outcomes of action. Renunciation of act and its outcomes is not the physical renunciation of them successively, but by being aware of them from the state of an inclusive and sensitive mind. According to the *Gita*, renunciation is a state of awareness. Hence, the *Gita* (4.19) emphasizes that he whose engagements are all devoid of desires and imaginations and whose actions have been burnt by the fire of wisdom, him the wise call a sage. Renunciation of judgments is the quality of a free mind. A free mind is a pure mind that is not swayed by dualities of giving up and possessing. For an inclusive mind, giving up is not a negative act but a positive state of being. In this sense, by "giving up" one gains and does not lose something. The spiritual gain is described by the *Gita* (4.24) as—*brahma karma Samadhi*—a meditative state of being which is neither gained nor displaced by a mental or physical act. Hence, for a true renunciant "pleasure and pain are same," and a truly renounced mind will be "free from the pairs of opposites." Renunciation refers not to the act itself but to the frame of mind behind the act. Therefore, with the absence of such a state of being, of true "giving up," even a supposedly altruistic act will not give the taste of true freedom anticipated from such an act.

In the *Bhagavad Gita* dialogue, there are instances where Krishna and Arjuna, the warrior hero, get into a discussion of complex processes that shape the connections among desire, action, and renunciation. The philosophy of action and renunciation or, in other words, possessing and giving up is centered on two questions: "What exactly is given up?" and "Why is it given up?" These questions lead us to the perception that renunciation is a state of being and not necessarily an act or idea. The idea of "giving up," in the *Gita*, has also provoked several controversial interpretations, the most popular of which is that renunciation is yet another physical act, of giving up objects, people, and relations. This misinterpretation is promoted by a hasty philosophy that "everything is illusory." The fact of the matter is that even physical giving up is not a difficult affair. What is challenging is to give up the feeling that "*I* have given up," which is the agent that is at the root of the mental intention or the physical act. "Giving up," according to the *Gita*, is neither a mere physical act or mental intention nor the abjuration of a social role. Renunciation is the ontological core of a person. Altruism is not an emotion or an action. It is the inclusive space defined by the core self.

9.7 Conclusion

In the mainstream discussion of altruism and altruistic acts, the popular trend is to classify certain behaviors as selfish and certain others as altruistic. The excessive interest in interpreting human acts in terms of behavior owes its origin to evolutionary theories where all acts are judged through one filter—acts that favor survival and acts that do not. If we follow such a divide based on behaviors, the appropriate classification would be selfish acts as behaviors and altruistic expressions as self-expressions. While altruism and selfishness are still being discussed using exclusive biological parameters (expectation for future reward, or avoidance of future punishment), our day-to-day life experiences tell us that an altruistic gesture, or an act, is embedded in self-space, with indicators such as love, purpose-perception, and inclusivity. Altruism is not an isolated orientation per se.

The labeling of an act or gesture as altruistic, or not, is mostly done by another individual. For the person who expresses altruism, his or her gestures and acts are just signposts of contentment that is experienced in the inner depths, and for that reason, there is no giving up or sacrifice, and no expectation or disillusionment, but only expressions of joy.

References

Bodhananda, S. (2005). *Psychology of the Gita*. Retrieved January 27, 2012, from Sambodh Foundation http://www.sambodh.org/September2004/messages/gita_psychology.htm.

Davis, R. H. (2005). Altruism in classical Hinduism. In J. Neusner & B. Chilton (Eds.), *Altruism in world religions* (pp. 159–178). Washington, DC: Georgetown University Press.

Dawkins, R. (1976). *The selfish sene*. Oxford: Oxford University Press.

Iverach, J. (1994). Altruism. In *Encyclopaedia of religion and ethics* (J. Hastings, Trans., Vol. 1, pp. 354–358). Edinburgh: T&T Clark.

Menon, S. (2002). The selfish meme and the selfless Atma. *Sophia, 41*(1), 83–88.

Menon, S. (2007). Basics of "spiritual altruism.". *The Journal of Transpersonal Psychology, 39*(2), 137–152.

Radhakrishnan, S. (1977). *The Bhagavad Gita* (Vol. 6th Indian. rpt.). Bombay: Blackie and Son (India).

Subramaniam, K. (1990). *Mahabharata*. Bombay: Bharatiya Vidya Bhavan.

Welsh, R., & Knabb, J. (2009). Renunciation of the self in psychotherapy. *Mental Health, Religion and Culture, 12*(4), 401–414.

Chapter 10
Spiritual Transformation and Healing: Is Altruism Integral?

Joan Koss-Chioino

10.1 Introduction

Explanations of spirit healing in the literature, although interesting, do not fully account for the spirit healers' work, the spirits' diagnoses and predictions during the healing ritual, or the affective expressions of healers and supplicants. The model of the ritual healing process I describe below (and elsewhere; see Koss-Chioino, 2006) focuses on spiritual transformation and empathy as central components. Its formulation has led some people to question the motivations of spirit healers in their lifelong work with supplicants. Smith, Lapinski, Bresnahan, and Smith (2013) in Chap. 2 of this volume describe five components of altruistic behavior as "… whether the act was motivated by a primary concern for the other, whether the actor would be likely to engage in self blame or be socially censured if he or she did not engage in the action, and the benefits to the recipient. The other two components of altruistic behavior are postulated to operate pan-culturally. These components are: empathy, and cost to the initiator." In my formulation of spirit healing process and altruism, all of these components are relevant, with much less emphasis and relevance of self-blame/social censorship. Empathy and cost to the initiator might be considered primary components as will be described below.

In many observations and over years of participation, it has become clear that a deeply caring attitude, although not always overt, is strongly associated with the development and practice of a spirit healer. Given the distress that healers experience on behalf of their supplicants (see below) and the spiritual and psychological benefits they often receive as part of their healing vocation, can they be said to be altruistic? The answer is a qualified "yes"—but how do they acquire that attribute?

J. Koss-Chioino (✉)
School of Evolution and Social Change, Arizona State University, Tempe, AZ, USA

George Washington University, Washington, DC, USA
e-mail: joan.koss@asu.edu

D.A. Vakoch (ed.), *Altruism in Cross-Cultural Perspective*,
International and Cultural Psychology, DOI 10.1007/978-1-4614-6952-0_10,

Empathy and altruism have been linked in a number of experimental studies (Batson, 2002; Batson, Ahmad, Lishner, & Tsang, 2002; Mikulincer & Shaver, 2005). A few have demonstrated that altruism is a core component of spirituality, a notion reflected in common discourse and many descriptive studies (Greenwald & Harder, 2003). Of interest to the present discussion is a study by Schwartz, Meisenhelder, Ma, and Reed (2003) which showed that helping others was associated with better physical and mental health among a stratified, random sample of members of the Presbyterian Church in the United States. Both helping others and receiving help were predictors of mental health, but giving help was a better predictor of mental health than receiving it. The literature that examines altruism and empathy is not well developed, but the studies which are available support the notion that the emergence of a deeply felt empathy (which is, at its fullest extent, the experience and practice of what I call "radical empathy") develops from the initial stages of spiritual transformation and leads healers to systematically respond altruistically to persons in distress who are seeking their help. The healers might be said to be volunteering in an altruistic way within a cultural context in which helping others is highly valued (see Chap. 6 of this volume by Grönlund (2013)).

In this chapter, I describe a foundational model of the ritual healing process based on my experiences of spirit healing in Puerto Rico—mostly Spiritist practices—in the United States, and elsewhere, as well as on the large volume of anthropological literature on ritual healing. I also consult the psychology literature on psychotherapy for possible parallels. I will then show how empathy develops in novitiate healers and becomes central to ritual healing and describe its role in healing process. Finally, I raise a question whether healers are altruistic in practicing their avocation. This is a complex question because once new healers have undergone the spiritual transformation that initiates their development as a medium, with the ability to communicate with the spirit world, their lives and well-being most often center around their work at a healing center and their spontaneous responses to distressed persons. Even though they are enjoined by their guardian spirits not to profit monetarily from healing work, the question of motivation complicates the picture since spirit healers feel compelled to carry out healing work for their own as well as others' well-being. Nevertheless, as I will show, the healer makes some significant sacrifices to do so.

10.2 Spiritism: A Popular, Recent Religion

Spiritism is a philosophical and scientific movement codified by a French scholar in Paris in the mid-nineteenth century. It developed from an exploration of disincarnate, eternal spirits that survive the death of human beings and is based on records of the experiences of young mediums in séances and collections of similar data from many groups around the world (Kardec, 1886). The scholar Hippolyte Denizard Rivail (1804–1869) wrote under the pseudonym of Allan Kardec and published six books and a review organ which have been (and continue to be) translated and republished in many countries. These works codified the practices and philosophy of Spiritism.

Spiritism's basic principles were borrowed from different philosophical, cultural, and religious traditions. Ideas about the Supreme Being, the life as eternal and spiritual and as an evolutionary experience, the immortality of the soul, the spirit communication, the law of cause and effect, and the reincarnation are not new. However, Spiritism contributes a new theoretical approach and an understanding of these ideas from rationalist, naturalist, and humanistic perspectives. The idea of a Supreme Being or God is, in Spiritism, the acknowledgment of the existence of an intelligence greater than an organized life. Life is inexorably without end; the spirit continuously progresses through different experiential stages of existence that guide it toward greater levels of spiritual maturity, individuality, and humanism (Denis, 1909; Novich-Hernandez, 1999). Therefore, the spirit/Self or consciousness, which is an emotional and intellectual entity, develops from being unconscious to conscious of its self-individuality and independent decision-making processes (Geley, 1995; Guimaraes-Andrade, 1992). The dynamics of this process is guided by free-willed decisions which generate positive and negative outcomes that influence the soul's evolutionary process. The spirit utilizes reincarnation as a method to access corporal life experiences, as well as those of a non-corporal nature. While in the spirit dimension, each entity assesses its moral and intellectual evolutionary necessities in order to select an appropriate physical life environment that cultivates and complements the continuous formation of the Self. In sum, life is viewed essentially as one, where sometimes the spirit acquires schooling through corporal bodies and, at other times, does this without a body.

The core Spiritist belief holds that there is life after the death of the body. This means that the disincarnated person's spirit can communicate with incarnated spirits after disposing of its body. The return of a spirit or its consciousness may cause positive or negative effects on the environment where such intelligent energy communicates. Communication will depend on the level of spiritualization which each spirit has attained within the spiritual levels. Materialistic spirits tend to focus their psychological necessities on the material world; these tend to deny their new spiritual reality and therefore insist on reenacting activities they performed while incarnated. On the other hand, spiritualized beings may reappear among the incarnated to share some idea that may inspire the people they observe to evolve on intellectual and moral terms. Therefore, Spiritist sessions continuously assess each of their participants' spiritual needs. And if a spirit is detected that insists on being intrusive, persistent, and recurrent over the free will of another person, mediums act to correct this situation, which they describe as "spirit obsession."

Spiritism postulates that neither demons nor evil exists; instead, life is surrounded by independent, self-evolutionary, decision-making processes that develop positive and negative effects toward the self and its environment. People and spirits are neither good nor bad; their moral character depends on their spiritual, intellectual, and moral self-evolution. Concerning mediumship, Aizpúrua (2000, p. 164) notes: "According to Spiritism, it should always be practiced towards goodness, taking advantage of an incomparable method that stimulates spiritual education, self-knowledge, information about life, and the practice of solidarity and fraternity."

Spiritism is not a religion in the conventional sense of an institutional organization or movement. Nevertheless, it is a religion from a philosophical and phenomenological perspective that seeks to unite humanity in acknowledging the self as a spiritual being (Kardec, 1979).

10.3 Core Elements of Ritual Healing Process

We find many rich, parallel descriptions of healing rituals that employ spirits, gods, or other extraordinary beings across cultures, and that explains why it is important to explore core elements of healing process. This is especially important if we assume that cultural elaborations—including different mythic worlds, symbol systems, and schemas to identify illness and disorder, as well as various types of ritual paraphernalia—are elaborations of content rather than process. This appears particularly true for spirit healing, which has very similar ritual forms, enhanced by local content, across diverse regions of the world. Some of what I describe here as foundational aspects of the ritual healing process have also been identified in some psychotherapeutic modalities, such as Jungian analysis and client-centered therapies, as important though not always essential to the therapeutic process. My formulation of the core components of the ritual healing process, using spirit healing as a prototype, focuses on the nascent healer's experience of spiritual transformation that often is associated with a severe illness or period of distress and an emerging ability to commune with the sacred, however conceived in diverse cultures, including as spirits, gods, saints or God, or another numinous being. This type of spiritual transformation generates a capacity for what I call "radical empathy" in the developing healer which shapes her healing work for the rest of her life. I describe a model of ritual healing, based on these foundational components, in the sections to follow.

10.4 Spiritual Transformation

Spiritual transformation, which can be defined as "Profound changes in the way people understand, approach, and experience whatever they hold as sacred" (Pargament, 2006), is the foundation of the ritual healing process. We might define it more specifically as "dramatic changes in world and self views, purposes, religious beliefs, attitudes or behavior. These changes are often linked to discrete experiences that can occur gradually or over relatively short periods of time" (Katz, 2004, p. 1). Spiritual transformation is reported as the hallmark of healer initiation by many spirit healers across cultures and also by some of their clients. (See, e.g., Csordas & Lewton, 1998; Katz & Wexler, 1990; Katz, 1993; Peters, 1981). It appears to be a central component of the healing process in many healing systems, with the exception of most Western biomedicine as well as some cosmopolitan healing systems, such as classical Chinese medicine, which are grounded in biophysical concepts. Furthermore, the bodily and/or psychic incorporation of spirits or God(s) or other extraordinary beings, as outcomes of a spiritual transformation, are directly associated with being healed, whether or not the sufferer's symptoms remain (Csordas, 1994; Kleinman & Sung, 1979). The role and effect of spiritual transformation on healers and healing process have been described in reports of ritual healing in a number of societies and groups (e.g., Barnes & Sered, 2005; Csordas, 1994; Katz, 1993; Koss-Chioino, 1992; Moodley & West, 2005; among many others).

10.5 The Wounded Healer

The anthropological literature contains numerous examples of indigenous healers and shamans who are initiated into their healing roles as the result of a serious, often life-threatening illness that is resolved when an extraordinary being(s) is introduced into their consciousness and life. Behavioral and attitudinal changes follow once the spirit becomes an integral part of the novitiate's life and being. Initiation into the healer role is both preceded and accompanied by changes in self- and worldviews. Descriptions of healer initiation following a life-threatening illness have lead to the formulation of the "wounded healer," whose source of power and authority as a healer is a continuous relationship to her woundedness, as exemplified by the healer's willingness to remain aware of those parts of herself that are perpetually wounded. Whether the healer's wounds are physical, emotional, or existential, they often must be confronted by the healer during transactions with ill and distressed clients.

The idea of the wounded healer continually repeating the process of being healed is not new. Apart from the anthropological literature on shamanism, Jung viewed the wounded healer as a central archetype represented by Asklepios, the God-healer and founder of Greco-Roman healing cults. Jung and some post-Jungians view this archetype as a key aspect of the analyst–patient relationship, one that activates the endogenous healer in the patient (Groesbeck, 1975; Guggenbuhl-Craig, 1978; see Kirmayer, 2003 for a detailed discussion of the wounded physician/healer; also Miller & Baldwin, 2000). In the Jungian formulation, as explained by Groesbeck (1975), the patient projects his inner healer onto the analyst because the distress of the wound blocks the inner healer. If the analyst does not recognize the healer–wound polarity in himself, he may project his woundedness onto the patient, which prevents healing. The patient's wound can be successfully treated when interaction with the analyst, who recognizes his own woundedness, facilitates the patient's withdrawing his projection of "healer" from the analyst so that the patient's own inner healer is activated.

In his cogent discussion of the wounded healer, Kirmayer (2003) cites Van Franz (1975), who relates Jung's work to that of "shamans and medicine men" who mediate between the spirit world and the suffering client. A key aspect is the way in which shamanic healers gain power and credibility through their inner experience of severe illness and recovery. During this event, most often with the intervention of other spirit healers, the novitiate develops the capacity (Spiritists use the word "faculty") to communicate with and control spirits or other extraordinary influences that can cause illness or distress. She also acquires one or more spirit guide–protectors to facilitate a new healing vocation.

10.6 A Case Example

In spirit healing in Puerto Rico, which I use as a prototype for this discussion, the openness to confront one's own woundedness (as related to partial recovery from a life-threatening illness) is commonly associated with new, life-shaking experiences of spirit beings who occupy a world parallel to that of living beings (see Hefner, 2006, for a description of the qualities of and encounters with the sacred). A woman in Puerto Rico recalled the events of her initiatory illness in great detail, although it occurred several years earlier. Excerpts from the transcript follow.[1]

Evelyn (a pseudonym) recalled:

> They gave me many remedies…but it seemed that nothing helped because I lacked faith… I was failing…One day the girl who had told me to go to an espiritista appeared at my house… The only thing she said was "you are still drinking water." I remember…that a loud shout came out of me of which I was aware, but nothing more. Well, they took me to an espiritista, but I was so wild, laughing at those things; well I believed that they had separated my body from my spirit…I didn't know where I was. ….In my ear I heard someone who said to me, "Look, now you are with me. You don't have a body because look where your body is." I didn't see anything…only a gray coffin with my body inside and someone was repeating many times, "Now you know that spirits exist. Look where your body is in that box. Do you see?"…I didn't have a body—nothing, nothing. In a moment I felt like my blood was circulating and my feet were feeling the ground but I tried to open my eyes and couldn't. At that moment, with my eyes closed, I saw a tall black woman with a dress woven of a very thick thread but she was smiling. (I guess that would be my *madama,* she interjected.)

The figure Evelyn saw is a spirit guide–protector of many women, the Caribbean obeah healer. Evelyn reacted with great fear at first because she believed that she was "really" dead. However, her husband brought a Spiritist medium to the house who worked with Evelyn, saying that she had just developed her "faculties" and he could see her house full of people seeking to help. After 2 weeks of rest, she went to a Spiritist center and was told that she had to do spiritual work and prepare a healing center at her home.

Communion with the sacred may take three main forms: visions of the spirit world, journeys to that world, and voluntary (involuntary for persons who are not "developed") embodiment (i.e., possession trance) by spirit beings. In Spiritism, for example, both protector–guide spirits and intrusive, often harmful, spirits may possess ("obsess") mediums; the harmful spirits displace the medium's own spirit.

[1] This is my translation.

Novitiate healers learn how to control communication with spirits or gods through the tutelage and/or observation of other healers. They then experience protector–guide spirits in ways that are both personally and cosmically meaningful.

Given the widespread belief in the special qualities of extraordinary beings across many cultures, one's experience of these beings during a life-threatening illness can make a significant impression upon the sufferer, who is in a state of high emotional arousal—confused, fearful, desperate, socially withdrawn, etc. As Frankl (1959) observed long ago (and as many writers have observed since), there are many reports of transcendence in the context of suffering, facilitated by both psychological and physical factors. If a healer is present who establishes a significant association between a spiritual entity and the hope of, or actual relief from, danger and/or suffering, the spirit can take on significant personal meaning for the sufferer. This then appears to reinforce or establish belief in the power of extraordinary beings (spirits, God, or gods as conceived in each group), as has been noted throughout years of anthropological writings on illness and healing. We can say, following Csordas (1994), that for those who hold a worldview centered on the self, the spirit-other becomes "embodied" within the self of healers. In Spiritism and numerous other spirit healing cults, embodiment of spirit is experienced as the continual, often lifelong, presence of one or more spirit guide-alters with whom a special relationship is maintained. A personal spirit protector–guide (or several) makes healing work not only possible but also safe from contagion, that is, immune to distress-causing spirit beings brought to the healing session by suffering clients.

Following such experiences, spiritual communion becomes the foundation of the healer's capacity for empathy. As I will describe in more detail later, the healer not only is empathic in her healing work but also comes to employ "radical empathy" in healing relationships constituted by visionary experiences, trance, and possession by spirits. Spirit work is based on the emergence of an intersubjective space where individual differences are melded into one field of feeling and experience shared by healer and sufferer. In Spiritism, intersubjectivity is essential to making a diagnosis (i.e., getting evidence) that describes the spirits and their reasons for causing distress to the sufferer.

10.7 Empathy

George H. Mead (1932) proposed that empathy is the ability to assume the attitude of another person. There is now a plethora of meanings and explications of "empathy" in psychology and other literatures. In the social psychological literature, for example, the concept may be referred to as "social insight," "interpersonal sensitivity," or "interpersonal judgment." In psychological research, empathy is the subject's ability to predict how another person will respond to items displaying certain psychological properties. Although empathy has been a focus of client-centered therapy (Rogers, 1957, 1959) and object-relations work by Kohut for decades, interest in its clinical role as a component of clinical process has expanded recently, going beyond the idea that empathy is useful only in establishing a therapeutic relationship as a "kindly and supportive posture" (Bohart & Greenberg, 1997, p. 4).

For Carl Rogers (1959, p. 210), the "state of being empathic" meant to "perceive the internal frame of reference of another with accuracy...as if one were the other person, but without ever losing the "as if" condition." Losing the "as if" meant a state of complete identification, which was to be avoided. Since the goal of client-centered therapy is to facilitate an authentic sense of self in the client, transiting personal boundaries is considered counter-therapeutic. Kohut (1984, p. 82) regarded empathy as "vicarious introspection" and said that having a similar experience allows us to gauge what another person is feeling. This is the methodological basis for analytic work. As a clinical tool, according to Rogers, empathy is the "capacity to think and feel oneself into the inner life of another person." Some of his followers have described "empathic attunement," or the "attempt to experience as closely as possible what the patient is experiencing" (Rowe & MacIsaac, 1989). Throughout these discussions, however, there is a thread of discomfort about the therapist's need to move in and out of an empathic state, because the psychotherapist must preserve a sense of neutrality.

Preston and Shumsky (2002, p. 48) describe an "empathic dance" in which "empathy is understood as a co-created web of meanings that are negotiated moment-to-moment, weaving the fabric of a new relational experience." The focus of the "dance" shifts from the analyst to the patient and then to the dyad as the therapy proceeds, but the main direction is the therapist's commitment to create a bridge to the patient.

Both the client-centered and self-psychology/psychoanalysis schools, among many others, subscribe to the ideal of the psychologically autonomous, integrated individual as normal (and preferable), and maintain that any dissolution of boundaries between persons is counter-therapeutic (O'Hara 1997). This also leads us to question if mutuality and genuine dialogue are possible in a therapeutic relationship, given the power differences between therapist and client (Kitschenbaum & Demanchick, 2004). O'Hara refers to Shweder and Bourne's (1982) description of the Western modernist self as based on an egocentric worldview in which the "person" (the inner self) is treated as if it lacks a social and cultural context, in contrast to persons who possess a sociocentric worldview in non-modernist societies. She asserts that modernist, "Western" psychology limits and shapes the use and meaning of empathy in therapeutic interactions. This perspective is directly relevant to the discussion of radical empathy to be developed below. Much ritual healing is embedded in a pervasive sociocentric worldview in which persons are fully integrated into primary groups in different social and cosmic realms.

10.8 Empathy in Ritual Healing

The ritual healing process in Spiritism (and similar rituals that heal with spirits) includes an empathic exploration in which healer-mediums both see and feel the sufferer's distress through the agency of spirits. There are three types of extraordinary states (e.g., altered states of consciousness) routinely present at the healing sessions: At first, the mediums enter into a quasi-trance state in which they call their guide–protectors to come to the table. These spirits "descend," possess them briefly, and then stand behind them while the healer-mediums work at the table on behalf of the assembled supplicants. As each supplicant is worked on, one medium will enter into direct contact with the distress-causing spirit attached to that person by "seeing" into the spirit world. That medium or another one then becomes deeply and often unconsciously immersed in the inner experience of the sufferer by "taking the spirit" into their body. The medium becomes possessed, and the spirit which is molesting the supplicant will then speak to the supplicant through the medium.

Spirit-mediums report that they feel a great deal of tension during spirit contacts: an "electric charge" that starts in the fingertips and goes through the body, as well as an accelerated "heart rate that can be very loud or violent." Mediums who have developed the faculty to take the illness-causing spirits into their bodies and become possessed by them may experience the spirit's feelings of anger and aggression toward the sufferer. Most often, the experience of possession by a distress-causing spirit is reported to be extremely unpleasant, even though healers say that they are not conscious of actual events that occur when they are possessed by a spirit. They do report feeling intense sensations of heat, sweaty hands, pain in their extremities, trembling, headaches, buzzing in their ears, hot and cold sensations deep within their bodies, high blood pressure, and feeling as if an electric current is passing through their heads or bodies. In direct contrast, when their protector–guide spirits come through them (as at the beginning of the session), they report mostly pleasant sensations. It might be noted that mediums are enjoined not to work at the table if they are ill or fatigued, as they will be more vulnerable to the influence of illness-causing spirits. Moreover, they are instructed to prepare for the session by eating little or no food, meditating, and relaxing in order to prepare their bodies for the difficult takeovers by spirits.

The spirit healing process completely sidesteps the concerns of many psychotherapists and analysts about the negative effect of mutuality in the therapeutic dyad, when empathy in treatment means attunement, sharing, or even resonating with the inner experiences of the client. The medium is only an intermediary for spirits; the drama of the sufferer's inner life does not touch her personally except as a call to use her faculties to contact and communicate with the spirit world (and the difficulties associated with that use). However, distress and suffering are calls to which she must respond even outside of the ritual session, if dictated to do so by her protector–guides, on whom her own well-being depends. Moreover, healer-mediums often express and act upon a collective sentiment central to their version of Spiritism (and a traditional sentiment in Puerto Rican culture that may be changing) that a person is continually affected by what other persons are feeling, particularly within families (Koss-Chioino, 1990). It must be further noted that the mediums at all ritual sessions refer to individuals as small units of the universe ("grains of sand"). And their spirit guides preach the need for all people to contribute to the universal "progress" of all spirits (i.e., to achieve peace, harmony, and balance) by exercising *agape* love.

10.9 Radical Empathy: A Step Beyond Empathy

"Radical empathy" takes empathic behavior further in that the wounded healer enters into the feelings of suffering and distress of people who attend the sessions and whom a spirit indicates need help (or, at rare times, persons they may meet in the course of their life). The healer-medium experiences feelings felt by the sufferer (*plasmaciones*), communicated through spirit visions (*videncias*) and/or possession by a spirit. Importantly, she has the guidance and authority of her spirit guide–protectors, who prevent her from being overwhelmed or seriously affected by the client's suffering. When a healer's own well-being and continued healing avocation depends upon a spiritual connection, the interpersonal space in which healing takes place becomes a sacred space and radical empathy acts as a path to transcendence by the assembled group.

We can examine radical empathy from a number of diverse perspectives. Spezio (2006) notes the widespread variation in the definition of empathy and observes that for some, such as Scheler (1954), "sympathy" is the more accurate term. Spezio's use of "sympathic" is similar to the notion of radical empathy that I develop here because it includes not only experiencing what another person feels but also acting upon that experience. Winnicott (1971) suggests that the patient and analyst coalesce psychologically in an "area of illusion" which is the place in which the analyst meets the psychic reality of the patient via her own psychic reality. Speaking from the perspective of analytical psychology, Samuels (1985, p. 52) describes a process found in Jungian analysis as an "embodied transference," which he describes as a "physical, actual, material, sensual expression in the analyst of something in the patient's inner world, a drawing together and solidification of this, an incarnation by the analyst of a part of the patient's psyche...." As we have already seen, in Spiritist healing, the sufferer's inner state is mirrored by the medium-healer, who reports that she feels the same pain, distress, or confusion as the sufferer with whom she is working. These feelings often come on unbidden, especially in novice mediums. A vivid example is that of a research assistant working in one of my projects who had been told by several Spiritist healers that she was developing into a medium:

> Sarah accompanied doña Maria on a visit to a bed-ridden client who had been paralyzed for four years. Sarah reported that at first she felt "deeply sorry for this woman." Shortly afterwards she "felt a creeping heaviness in her arm which traveled down her spine to the middle." Then she got the same feeling in her legs—especially one leg. Suddenly she was unable to move her legs. The client reported that the paralysis came upon her in exactly the same way over four days. She had never told anyone exactly how it happened. Doña Maria told Sarah that the feelings would leave if she described them verbally to the client and they did. Sarah then felt "a calmness inside."

Samuels (1985, pp. 58–59) talks about embodied transference in analytic work as the sharing of a mythic world, a *mundus imaginalis*, which "refers to a precise order or level of reality," "an intermediate dimension...in between patient and analyst," in between body and mind, and in between the analyst's conscious and unconscious. It is a world shaped by the analyst–patient relationship, "imaginally but not subjectively real." While these concepts and terms are mainly meaningful to analytical psychologists, who work with a model of a therapeutic dyad in which the separateness of the individual actors may be transcended by an "embodied transference," as described by Samuels, the parallels to Spiritist notions and spirit healing process are striking. For people who believe that spirits exist in a parallel world, there is a shared imagery of spirit phenomena (Koss-Chioino, 1996). In contrast to the dyadic analytic situation described by Samuels, the imaginal world in spirit healing is shaped by a three-party relationship among the sufferer, the medium-healer (or mediums) who are the sufferer's conduit to the spirit world, and the particular spirit or spirits brought to the healing table. It is also shared, to lesser extent, by all who attend the sessions. This very different therapeutic structure has a distinct advantage for the spirit healer. She does not bear responsibility for the effect of spirits on a sufferer, either positive or negative, or for a cure; the spirit-medium does not herself "heal." Despite a strict code of ethics in Spiritism, there is no need for healer-mediums to fear that they have transgressed personal boundaries (a concern for therapists), since it is the spirits who invade personal space and the healers who endeavor to "take off" (*despohar*) these invaders using the faculties given them by their spirit guide–protectors.

10.10 Altruism and Ritual Healing

The concept of altruism is rather new; the word was coined by Auguste Comte (1798–1857) "to displace terms burdened by a theological history" (Post, Underwood, Schloss, & Hurlbut, 2002, p. 3). It relates closely to "love" as *agape*, which conveys a sense of sacred rather than romantic love. In common usage, "altruism" refers to helping another being without expectation of benefit to oneself. Kristeller and Johnson (2005, p. 393), who focused on relationships among altruism, empathy, and meditation, suggest with Post et al. (2002) that self-transcendence is "a necessary precursor to altruism" and compare this approach to the Buddhist concept of loving-kindness. They note the universality of these concepts (in many religions), phrased in Buddhism, for example, as compassion, an association between "suspending a sense of self (concern)" and relieving suffering in others. An outcome of spiritual transformation in the spirit healers I studied is the appearance of feelings and practices indicating new and deep connections to others, particularly in the sense of a cosmic connection to all human beings, past and present. However, this feeling of connection does not seem to result in the complete "suspension of self-engagement," which is proposed by Kristeller and Johnson (2005, p. 394) to be an essential aspect of altruism (and empathic behavior). When ritual healers are entranced—that is, possessed by either their protector–guide spirits or the spirit *causas* of clients' distress, as in spirit healing practice, for example, there is a clear "suspension of self-engagement." What seems relevant to ritual healers—who are healing themselves at the same time that they heal others—is that their motivations are mixed. It is difficult in this context to speak of a concept of "motivations" as altruistic (or not) in that the healers feel compelled to act as they do because their spirit guides mandate it for reasons discussed above.

An interesting perspective on the spirit healers comes from studies by Mikulincer and Shaver (2005, p. 34) who examine the relationship between attachment security and altruism. They found that compassionate feelings and altruistic behaviors "are promoted by both dispositional and experimentally induced attachment security." In the case of the spirit healers, the process of spiritual transformation usually brings the novitiate healer out of a state of isolation related to recovering from a severe illness or distress; it then leads to a number of new attachments; to a healer-mentor, a group of spirit healers who ally with the novitiate healer as she begins to work with them; and to supportive, spirit guide–protectors. An enhanced sense of connection to other beings that emerges very clearly—in reports of their experiences of spiritual transformation—changes feelings of isolation and abandonment. Mikulincer and Shaver (2005) conclude that more precise studies need to be carried out; however, their ideas suggest that a healer's work acts to alleviate attachment insecurity by extending feelings of connection and of feeling continually needed.

This type of inquiry depends on how the parameters of altruism are defined. Batson and his colleagues (Batson et al., 2002) define altruism as a specific type of motivation when one's goal in helping another person is to increase their welfare. Other definitions are more exacting: there is no expectation of gain for self on the part of the altruist, thus excluding any self-interest. Moreover, the intention must be to benefit another person; the altruistic act cannot not be a secondary consequence (Renwick Monroe, 2002). Oliner (2002) asserts that the altruistic act must inflict risk or sacrifice on the altruist.

Considering Spiritist healer-mediums, one prominent Spiritist commented on their moral practice: "One must do good for the sake of good itself" (Aizpúrua, 2000, p. 274). This perspective motivates Spiritists to scrutinize the sufferers' moral/ethical limitations in order to understand how these affect the sufferer's current life and spiritual evolution. They then help the sufferer become conscious of such moral issues in order to overcome them.

While the mediums' behaviors show that they are very caring and moral persons, these behaviors are not, in my opinion, what constitutes "altruism" in spirit healing. According to Post et al. (2002, p. 3) "altruism" is "other regarded"-ness and "generous self-giving." In this sense, I propose that spirit healing (and most ritual healing) is based on altruism: Healers exert a great deal of effort, a large part of their life and time, and even risk their bodies, routinely enduring a good deal of discomfort to help others in distress. Whether this is inner-directed or indicates "self-transcendence" seems beside the point. As reported by many altruistic persons (Colby & Damon, 1992), ritual healers feel they have no choice but to carry out the healing work which was thrust on them at the time of their initiation and which they feel they must always continue. It is also through this work that they maintain their own well-being by acting as agents who restore well-being to others. We might ask: is selflessness aligned with self-interest, as in this example, altruism?

10.11 Conclusion

Radical empathy and spiritual transformation, as I have defined and described them, take a healer across a wide and deep emotional spectrum that relatively few psychotherapists or medical doctors—apart perhaps from some psychoanalysts—would welcome (Groesbeck, 1975). It appears that some persons who have the beliefs, the emotional flexibility, and the courage to deal with alterations in consciousness, as well as with the fascinating but often fearsome sacred realms (see Hefner, 2006), become ritual healers. The belief systems of ritual healers often provide ways to deal with emotional intensity and the impact of intense feelings of distress coming from a client, through both depersonalization (i.e., the spirits/gods/God heal, not the healer herself) and belief in mythic structures that explain these experiences. In Spiritism, rules about how to deal with vulnerability and to avoid contagion from the illness-bearing, spirit-ridden supplicants are transmitted during the informal tutelage that follows the initiation experience of spiritual transformation so that some sense of security and emotional support are provided. In addition, spirit healers most often work in teams, using their personal spirit protectors to support each other when distress-causing spirits descend upon the medium-healer during the session. It seems clear that the healers who develop the capacity for radical empathy and take the distress-causing spirits into their bodies on behalf of supplicants, or release their own spirits to travel to the world of nonliving beings in order to heal other persons, undergo a good deal of discomfort—especially when they are inexperienced at doing this. I suggest that suffering while seeking to increase the well-being of others qualifies spirit healers as altruistic; they consistently exercise other-caring behaviors. Many accounts of altruism suggest that such acts are performed without undue concern for oneself.

Acknowledgments This chapter is an adaptation of Koss-Chioino (2006). Spiritual transformation, ritual healing, and altruism. *Zygon, 41*(4), 877–892. It is republished in this volume with permission.

References

Aizpúrua, J. (2000). *Los fundamentos del espiritismo*. Caracas, Venezuela: Ediciones CIMA.

Barnes, L. L., & Sered, S. S. (2005). *Religion and healing in America*. Oxford: Oxford University Press.

Batson, C. D. (2002). Addressing the altruism question experimentally. In S. Post, L. G. Underwood, J. P. Schloss, & W. B. Hurlbut (Eds.), *Altruism and altruistic love: Science, philosophy and religion in dialogue* (pp. 89–105). New York: Oxford University Press.

Batson, C. D., Ahmad, N., Lishner, D. A., & Tsang, J. (2002). Empathy and altruism. In C. R. Snyder & S. J. Lopez (Eds.), *Handbook of positive psychology* (pp. 485–498). New York: Oxford University Press.

Bohart, A. C., & Greenberg, L. S. (1997). *Empathy reconsidered: New directions in psychotherapy*. Washington, DC: American Psychological Association.

Colby, A., & Damon, W. (1992). *Some do care: Contemporary lives of moral commitment*. New York: Free Press.
Csordas, T. (1994). *The sacred self: A cultural phenomenology of charismatic healing*. Berkeley, CA: University of California Press.
Csordas, T., & Lewton, E. (1998). Practice, performance and experience in ritual healing. *Transcultural Psychiatry, 35*, 435–512.
Denis, L. (1909). *Here and hereafter: Being a treatise on spiritual philosophy*. New York: Brentano's.
Frankl, V. E. (1959). *Man's search for meaning*. Boston, MA: Beacon Press.
Geley, G. (1995). *Del inconsciente al consciente* [Unconscious to the conscious]. Caracas, Venezuela: Ediciones CIMA.
Greenwald, D. F., & Harder, D. W. (2003). The dimensions of spirituality. *Psychological Reports, 92*(3), 975–980.
Groesbeck, C. J. (1975). The archetypal image of the wounded healer. *The Journal of Analytical Psychology, 20*(2), 122–145.
Grönlund, H. (2013). Cultural values and volunteering: A cross-cultural perspective. In D. A. Vakoch (Ed.), *Altruism in cross-cultural perspective*. New York: Springer.
Guggenbuhl-Craig, A. (1978). *Power in the helping professions*. Dallas, TX: Spring.
Guimaraes-Andrade, H. (1992). *Muerte, renacimiento, evolution* [Death, rebirth, evolution]. Caracas, Venezuela: Ediciones CIMA.
Hefner, P. (2006). Spiritual transformation and healing: An encounter with the sacred. In J. Koss-Chioino & P. Hefner (Eds.), *Spiritual transformation and healing: Anthropological, theological, neuroscience and clinical perspectives*. Lanham, MD: Altamira Press.
Kardec, A. (1979). *Viaje espirita en 1862*. Buenos Aires: Editora Argentina.
Kardec, A. (1886). *The spirits book*. Boston: Colby and Rich.
Katz, R. (1993). *The straight path: A story of healing and transformation in Fiji*. Reading, MA: Addison-Wesley.
Katz, S. H. (2004). *The spiritual transformation scientific research program*. Philadelphia, PA: Metanexus Institute on Religion and Science.
Katz, R., & Wexler, A. (1990). Healing: A transformational model. In K. Peltzer and P. Endigbe (Eds.), *Clinical Psychology in Africa*. Eschborn, Germany: Fachbuchhandlung für Psychologie.
Kirmayer, L. J. (2003). Asklepian dreams: The ethos of the wounded-healer in the clinical encounter. *Transcultural Psychiatry, 40*, 248–277.
Kirshenbaum, H., & Demanchick, S. (2004). Much ado about dialogue. [Review of K. Cissna & R. Anderson, *Moments of meeting: Buber, Rogers and the potential for public dialogue*.]. *Contemporary Psychology, 49*(4), 422–424.
Kleinman, A., & Sung, L. (1979). Why do indigenous practitioners successfully heal? *Social Science & Medicine, 13B*, 7–26.
Kohut, H. (1984). *How does analysis cure?* Chicago: University of Chicago Press.
Koss-Chioino, J. (1990). Somatization and somatic complaint syndromes among hispanics: Overview and ethnopsychological perspectives. *Transcultural Psychiatric Research Review, 27*(1), 5–29.
Koss-Chioino, J. (1992). *Women as healers, women as patients: Mental health care and traditional healing in Puerto Rico*. Boulder, CO: Westview Press.
Koss-Chioino, J. (1996). The experience of spirits: Ritual healing as transactions of emotion (Puerto Rico). In W. Andritzky (Ed.), *Yearbook of cross-cultural medicine and psychotherapy* (Ethnopsychotherapy, Vol. 1993, pp. 251–271). Berlin: Verlag für Wissenschaft und Bildung.
Koss-Chioino, J. (2006). Spiritual transformation, relation and radical empathy: Core components of ritual healing process. *Transcultural Psychiatry, 43*(4), 652–670.
Koss-Chioino, J., & Hefner, P. (Eds.). (2006). *Spiritual transformation and healing: Anthropological, theological, neuroscience and clinical perspectives*. Lanham, MD: Altamira Press.
Kristeller, J. I., & Johnson, T. (2005). Cultivating loving kindness: A two-stage model of the effects of meditation on empathy, compassion and altruism. *Zygon, 40*(2), 391–408.

Mead, G. H. (1932). *The philosophy of the present*. Chicago: University of Chicago Press.

Mikulincer, M., & Shaver, P. R. (2005). Attachment security, compassion and altruism. *Current Directions in Psychological Science, 14*(1), 34–38.

Miller, G. D., & Baldwin, D. C., Jr. (2000). Implications of the wounded healer paradigm for the use of self in therapy. In M. Baldwin (Ed.), *The use of self in therapy* (2nd ed., pp. 243–262). New York: Haworth Press.

Moodley, R., & West, W. (2005). *Integrating traditional healing into counseling and psychotherapy*. Thousand Oaks, CA: Sage.

Novich-Hernandez, H. (1999). *Salud, enfermedad y muerte* [Health, illness and death]. Caracas, Venezuela: Ediciones CIMA.

O'Hara, M. (1997). Relational empathy: Beyond modernist egocentrism to postmodern holistic contextualism. In A. C. Bohart & L. S. Greenberg (Eds.), *Empathy reconsidered* (pp. 295–319). Washington, DC: American Psychological Association.

Oliner, S. P. (2002). Extraordinary acts of ordinary people: Faces of heroism and altruism. In S. Post, L. G. Underwood, J. P. Schloss, & W. B. Hurlbut (Eds.), *Altruism and altruistic love: science, philosophy and religion in dialogue* (pp. 123–139). New York: Oxford University Press.

Pargament, K. (2006). The meaning of spiritual transformation. In J. D. Koss-Chioino & P. Hefner (Eds.), *Spiritual transformation and healing: Anthropological, theological, neuroscience and clinical perspectives*. Lanham, MD: Altamira Press.

Peters, L. (1981). An experiential study of Nepalese shamanism. *Journal of Transpersonal Psychology, 13*, 1–26.

Post, S., Underwood, L. G., Schloss, J. P., & Hurlbut, W. B. (Eds.). (2002). *Altruism and altruistic love: Science, philosophy and religion in dialogue*. New York: Oxford University Press.

Preston, L., & Shumsky, E. (2002). From an empathic stance to an empathic dance: Negotiation. In A. Goldberg (Ed.), *Progress in self psychology* (pp. 47–61). Hillsdale, NJ: Analytic Press.

Renwick Moore, K. (2002). Explicating altruism. In S. Post, L. G. Underwood, J. P. Schloss, & W. B. Hurlbut (Eds.), *Altruism and altruistic love: Science, philosophy and religion in dialogue* (pp. 106–122). New York: Oxford University Press.

Rogers, C. R. (1957). The necessary and sufficient conditions of therapeutic personality change. *Journal of Consulting and Clinical Psychology, 21*, 95–103.

Rogers, C. R. (1959). A theory of therapy, personality and interpersonal relationships, as developed in the client-centered framework. In S. Koch (Ed.), *Psychology: A study of a science* (pp. 184–256). New York: McGraw Hill.

Rowe, C., & MacIsaac, D. S. (1989). *Empathic attunement: The technique of psychoanalytic self psychology*. Northvale, NJ: Jason Aronson.

Samuels, A. (1985). Countertransference, the *mundus imaginalis* and a research project. *The Journal of Analytical Psychology, 30*, 47–71.

Scheler, M. (1954). *The nature of sympathy*. (P. Heath, Trans.). London: Routledge and Kegan Paul.

Schwartz, C., Meisenhelder, J. B., Ma, Y., & Reed, G. (2003). Altruistic social interest behaviors are associated with better mental health. *Psychosomatic Medicine, 65*(5), 778–785.

Shweder, R., & Bourne, E. J. (1982). Does the concept of person vary cross-culturally? In A. J. Marsella & G. M. White (Eds.), *Cultural conceptions of mental health and therapy* (pp. 97–137). Boston, MA: Kluver.

Smith, S. W., Lapinski, M. K., Bresnahan, M. J., & Smith, S. L. (2013). Conceptual aspects of altruism in cross-cultural contexts. In D. A. Vakoch (Ed.), *Altruism in cross-cultural perspective*. New York: Springer.

Spezio, M. (2006). Narrative in holistic healing: Empathy, sympathy, and simulation theory. In J. D. Koss-Chioino and P. Hefner (Eds.). *Spiritual transformation and healing: anthropological, theological, neuroscientific, and clinical perspectives* (pp. 206–222). Lanham, MD: AltaMira Press.

Van Franz, M. (1975). *C.G. Jung: His myth in our time*. Toronto: Little, Brown.

Winnicott, D. W. (1971). *Playing and reality*. London: Tavistock Publishers.

Chapter 11
Altruism in Human Ritual

Garry Chick

11.1 Introduction

> These good acts give pleasure, but how it happens that they give us pleasure? Because nature hath implanted in our breasts a love of others, a sense of duty to them, a moral instinct, in short, which prompts us irresistibly to feel and to succor their distresses.
> Thomas Jefferson, letter to Thomas Law Poplar Forest, June 13, 1814

Altruism is generally defined as the selfless concern for the well-being of others. In the case of nonhuman animals, that translates to behavior that appears to be detrimental to the survival of a given individual but which may contribute to the survival of the others. Calls by prey species that warn others of the approach of predators, for example, are often regarded as altruistic in that they may help the majority of animals survive while simultaneously drawing the attention of the predator to the individual giving the warning. While engaging in rituals rarely puts humans at risk from predators, aspects of many rituals appear to be largely selfless and may have negative repercussions for those who sponsor, manage, or support the events. Examples include many rites of passage, that is, ceremonies that celebrate the transition of an individual, or sometimes a group of individuals, from one life stage to another (Chick, 2004; van Gennep, 1960). Such rites are observed at events such as births, deaths, transitions from childhood to adulthood, marriages, religious affirmations and confirmations, retirements, graduations, initiation into secret societies (such as the Freemasons), and many others cross-culturally. These often involve gifting, feasting, and great ceremony, such as in elaborate weddings, debutante balls, or college graduations, often at great cost to parents, relatives, and others. Most of these appear to have altruistic aspects in that they involve not only substantial sacrifice by some but benefits to others.

G. Chick (✉)
Department of Recreation, Park, and Tourism Management,
The Pennsylvania State University, 801 Ford Building,
University Park, PA 16002, USA
e-mail: gchick@psu.edu

D.A. Vakoch (ed.), *Altruism in Cross-Cultural Perspective*,
International and Cultural Psychology, DOI 10.1007/978-1-4614-6952-0_11,

One example is the system of festivals in Latin America organized to venerate saints in the Catholic pantheon. In many traditional communities, festivals honoring locally important saints are sponsored, organized, and funded by community members. The classic system entails a series of hierarchically arranged sponsorship positions that involve both political and religious aspects of community life. Participation in the higher levels of the system can entail considerable cost in terms of both time and money for community members. But why do individuals participate in the *cargo* system? Various explanations have been offered, including that it acts as an economic leveling system (e.g., Nash, 1958). It may also function as a means by which villagers can acquire prestige (e.g., Cancian, 1965). Indeed, those who hold the highest *cargos* in their villages typically are accorded elder status with substantial decision-making power. But why would individuals voluntarily, and often eagerly, participate in a system designed to redistribute their wealth? And, of the many who participate in the system, only a few ever make it to the top, so prestige remains unevenly distributed. An alternative explanation is that *cargo* holding is an example of altruistic service to one's community, a contribution to the well-being of the group while having potential or real harmful effects for individual *cargo* holders. The purpose of this chapter is to examine evidence for the expression of altruism in the *cargo* system and to suggest that rituals, more generally, can involve altruistic behavior toward others.

11.2 Ritual, Expressive Culture, and the *Cargo* System

Human culture, defined as socially transmitted and shared information generally in the forms of beliefs and values, can be roughly divided into utilitarian and expressive components. The former deals primarily with how people go about making a living and raising a family, while the latter gives meaning to much of the former as well as to life in general. Anthropologists have devoted far more research attention to utilitarian than to expressive culture although particular activities, such as games, sports, art, music, and narratives such as legends, folktales, and other oral or written literature, have received considerable notice. Ritual is an anthropological favorite, and if defined as an organized and generally repetitive set of symbolic acts designed to communicate meaning, rituals are important and very common forms of expressive culture cross-culturally. Rituals can be both sacred and secular and used to remove sin, bring rain, grow crops, heal the sick, get politicians elected, make graduations memorable, unite couples in marriage, and get sports events underway, among others.

11.2.1 The Cargo *System*

Rituals, as part of the *cargo* (meaning "load" or "burden") or *fiesta* (meaning "festival") system, have occupied a place in Latin American expressive culture since the sixteenth century. The *cargo* system is an important religious (and now often secular) system that consists of a set of more or less hierarchically organized positions, generally held for the duration of 1 year, wherein community members sponsor and administer events based on the local religious calendar (Carrasco, 1961; Cancian, 1965; Chick, 1981, 1989, 2002; Dewalt, 1975). In some cases, individuals alternate between holding religious and secular political offices, while, in others, the religious and political systems are distinct. In San Rafael Tepatlaxco, a community of approximately 1,050 people in the state of Tlaxcala, Mexico, where I did field research from 1977 to 1980, the religious and the political systems were separate. I will describe the system, as it existed during my field research there.

The rule of thumb in San Rafael Tepatlaxco was that if there was an image of a saint in the community church, then a festival must be sponsored in the honor of that saint during the year. However, not all festivals were created equal. Some, such as the festival in honor of Saint Rafael, the patron of the village, were very elaborate (and expensive). The festival of Saint Rafael lasted for 2 weeks in October. The festival in honor of the Virgin of Guadalupe, the patron saint of Mexico, was also very elaborate and lasted for a week in December (the national celebration falls on December 12). These two festivals were each sponsored and organized by three men. The *mayordomo* was always an older, experienced hand at festival sponsorship and he had the assistance of two others, the *devoto* and the *topile*. Collectively, the group was referred to as *mayordomos* and the sponsorship itself, the *cargo*, was termed a *mayordomía*. Twelve other saints were honored with festivals but all of these were considerably less elaborate than those of Saint Rafael and the Virgin of Guadalupe. Two individuals, the *mayordomo* and the *devoto*, constituted the *mayordomos* for these smaller festivals, which ranged from a 3-day celebration for *la Vírgen Purísima* to nothing more than a dinner with a few guests for *San Gabriel*.

While, historically, and in many Mesoamerican and other Latin American communities, the *mayordomos* themselves supported the festival with their own money, in Tepatlaxco, most expenses were defrayed through donations collected from community members (and the visiting anthropology graduate student). So, while the *cargo* system has been described in the anthropological literature as a "leveling system" whereby resources are redistributed among community members from wealthier *cargo* sponsors (e.g., Nash, 1958; see Chap. 4 of this volume by Gibbons (2013), for a similar claim regarding Mayan communities), this was not an accurate description of Tepatlaxco because of the high level of support provided by villagers. The *mayordomos* did spend considerable time and effort on organizing the festivals, however. Cancian (1965) claimed that the *cargo* system acts as a stratifying mechanism that separates the community into multiple levels of social status through the award of prestige to *cargo* holders. He claimed, as well, that *cargo* holding legitimized differences in wealth that already existed, leading to community homeostasis and internal accord. This explanation is dubious for Tepatlaxco for, while wealth and status distinctions were evident, there was no obvious correlation between wealth and status in the community or between wealth and *cargo* service (Chick, 1980).

Why did villagers spend so much time and effort on the *cargo* system? It may have been an economic leveling system in some places in Latin America but not in

Tepatlaxco. The sponsorship of festivals in honor of saints certainly also had religious aspects. While community members in Tepatlaxco were nearly all Roman Catholics (there were two families of Protestants in the village, as well), in their version of Catholicism, the saints were intermediaries to God. Hence, propitiation of the saints was important. But village men fell into three groups, two of which participated in the system and one that did not, apparently due to different motives (Chick, 1989). The latter group consisted largely of men who were involved in daily, weekly, or longer period migrant labor. Individuals might commute (by bus) daily to nearby villages or towns or to the large city of Puebla, some 20 km distant. Others would commute to Mexico City or more distant cities on a weekly basis, while still others participated in transmigration to places such as Veracruz for agricultural work. Many of the older men of the village had participated in the *bracero* program in the 1950s and early 1960s and had been to places like Texas, Arizona, and Kansas but also to Illinois and Michigan (where they picked sugar beets). These men sometimes told me that they would like to participate in the *cargo* system but could not because of their work situation. Others, older individuals or those who were still participating, sometimes claimed that those who used their work as an excuse to avoid *cargo* service were simply lazy or not community minded.

Two groups of individuals did participate and the men in each of these could be distinguished by their patterns of *cargo* sponsorships. One group sponsored *cargos* in a pattern generally concordant with the hierarchical organization of the offices. That is, they began their *cargo* careers with relatively low-ranking, cheaper, and less onerous sponsorships and successively held higher-ranked positions until they completed the highest *cargo* in the village. They then became elders and wielded substantial decision-making authority, especially with respect to the *cargo* system itself. Individuals in the second group often sponsored as many cargos as those who went on to become elders but, instead of holding successively higher positions, they muddled around in the lower and middle levels of the system, sometimes holding higher offices, sometimes lower, and sometimes on the same level as their previous office. It could be argued that the nonparticipants and the elder status seekers were acting selfishly, but it would be very difficult to do that with respect to the muddlers.

Moreover, members of the two Protestant families in the village commonly participated in the *cargo* system despite the fact that the *mayordomías* were held to honor Catholic saints. Their participation had nothing to do with religious devotion; their participation was not going to save their souls, at least in their minds. A couple of the Protestant men told me that they participated because Tepatlaxco was their village, their home, and the others who lived there were members of their community and their friends. So their service in Catholic festivals (as well as in construction projects on the [Catholic] church in the village) made complete sense to them as a contribution to their community. In turn, the Protestant families were treated well, if occasionally with a bit of suspicion, and were regarded as integral members of the community.

Kurzban and Houser (2005) suggest that humans come in three types with respect to cooperation. First, there are cooperators, "who contribute to generating group benefits at a cost to self." Second, there are "free-riders, who do not incur these costs." Finally, there are "reciprocators, who respond to others' behavior by using a conditional strategy" (p. 1803). They experimented with multiple players, in groups, in a computer simulation wherein players allocated tokens to individual and group exchanges in order to gain points. Kurzban and Houser termed those who contributed little to the pooled exchanges "free-riders." Those who contributed a great deal most of the time were "cooperators," and those who contributed an amount that was about equal to that of others were "reciprocators" or "conditional cooperators" (Kurzban & Houser, p. 1803). The authors based their inferences about their informants' types from a plot of each player's contributions compared with the average contribution observed before making his or her contribution. Cooperator's contributions were well above the 45° line on the plot while free-rider's contributions were well below the line. Reciprocators' contributions clustered near the 45° line. Kurzban and Houser found 17 of their 84 informants (20 %) to be free-riders, 11 (13 %) to be cooperators, 53 (63 %) to be reciprocators, and three to be unclassifiable according to their criteria.

Kurzban and Houser (2005) feel that their findings support the idea of multiple and stable behavioral types that vary in terms of willingness to cooperate in group contexts. Their results corroborate others where researchers classified people as competitors (motivated to get better payoffs than others), cooperators (motivated to contribute to group welfare), and individualists (motivated to serve their own interests) (e.g., Komorita & Parkes, 1995) and experimental economics where people were found to be "spiteful" (competitors), "altruistic" (cooperators), and "payoff-maximizing" (individualists) (e.g., Cason, Saijo, Yamato, & Yokotani, 2004).

I suggest that these characterizations fit the three types of interactions with the *cargo* system that I observed in Tepatlaxco from 1977 to 1980. Those who completely avoided participation in the system might be termed "free-riders" as they often enjoyed the fruits of the system (i.e., the festivals and associated activities) but contributed little or nothing to it. Those whose "*cargo* careers" closely paralleled the hierarchical organization of the system and passed the highest office, thus becoming village elders, were individualists who maximized their payoffs. Those who I termed "muddlers" above participated intermittently and seemed to do so in order to contribute to the community. As such, they were cooperators or *altruists*.

11.2.2 A Cargo *Festival*

In 1978, the festival in honor of the patron saint of San Rafael Tepatlaxco, Saint Rafael, was held between October 19 and October 31. During this period, 15 Masses were celebrated, there was a *Mañanita* performance (the singing of traditional songs by young girls and boys), and residents carried images of the saints in two processions through the village. They were joined in this by people from several neighboring villages. Community members put up banners and streamers to decorate the community and bouquets of flowers were placed throughout the village.

Many of the activities that took place were religious in nature although it can be argued that religion, as an expressive system, has innate recreational qualities. Villagers, in addition to the *mayordomos*, served festival meals, almost always involving *mole*, a sauce with chocolate as a main ingredient usually used over meat. Many former villagers returned for the festival, which featured carnival-like games, baseball, soccer, and horse racing. There were several fireworks displays and vendors, set up around the church, sold ice cream, candy, and trinkets. Two dances were held in the elementary school courtyard. Alcoholic beverages were consumed in vast quantities (generally in the forms of beer, brandy, and *pulque*, the local beverage fermented from the sap of the *maguey* or agave cactus). The church itself was decorated with banners and, on the floor inside, with elaborate *alfrombras* (depictions of religious scenes on the church floor made from flower petals).

Other festivals ranged from an evening to a week in duration. Most villages in the area had between 15 and 30 festivals per year. The festivals associated with the patron saint of the community and the Holy Week were the most elaborate. The activities associated with the *cargo* festivals are the only form of community-wide recreation and are the only form of recreation to which all members of communities, whether young or old, male or female, are welcome. All other forms of recreation involve either individuals or small groups, including families. And, again, all of this is organized and, to a significant extent, financed by community members on a volunteer basis (see Chap. 6 of this volume by Grönlund (2013), on volunteering).

11.2.3 Research on Cargo *Careers*

I collected the *cargo* careers of 60 individuals from the local church records. The names of *cargo* holders were recorded each year, beginning in 1920 when Tepatlaxco achieved *Pueblo* status and then, in following years, held *mayordomías* separately from Santa Ana Chiautempan, the local municipal seat. The data were relatively complete through 1980 although the books for 3 years (1947, 1949, and 1951) were missing. I was able to partially complete the lists of officeholders for these years from the records of other years. Additionally, data for several of the lowest-ranking *cargos*, the *Mayordomía del Santo Entierro*, the *Mayordomía del Divino Rostro*, and the individual days of the Holy Week were often not recorded. Hence, I excluded these from the analyses described below.

Twenty-seven of the 60 individuals had achieved the status of elder in Tepatlaxco; that is, they had passed the highest office in the community, that of *fiscal* (the *fiscal* is the chief officer of the *fiscalía*, the lay governing body of all religious activities in the village). Based on the data available, these men had held between 5 and 11 (mean = 6.95, SD = 1.68) recorded offices. It is likely that each had held several more *cargos* that were unreported in the records. Of the remaining 33 individuals, two had reached the second highest level of the nine-level hierarchy (the *Mayordomo de San Rafael*), eight had reached the third highest level (*Mayordomo de la Vírgen de Guadalupe*), thirteen had reached the fourth highest, and ten the fifth highest. It should be pointed out that several individuals who had not reached the highest office still had the opportunity to do so. I did not gather information on anyone who had not held any offices.

So, 45 % of the 60 individuals had *cargo* careers that could be regarded as competitive; that is, they were motivated to surpass the efforts of others in the system, while 55 % were cooperators in that they seemed to participate in order to advance the welfare of the community. Since I was unable to determine the number of non-participants, the percent of "free-riders" cannot be calculated. While the 45–55 % ratio of competitors and cooperators seems to differ from 63 % to 13 % of reciprocators and cooperators that Kurzban and Houser (2005) found in their sample, the contexts these data represent were very different. Since I chose to ignore *cargo* careers of individuals who had held fewer than five offices, and since none of the competitors had held fewer than six offices, it is certain that more than 55 % of the individuals who participated at all in the system were cooperators, or *altruists*.

It is also possible that the competitors knew the system better than the cooperators did and were therefore better able to move through it systematically. Fortunately, this is a testable proposition. I asked a sample of 31 men, all who had had sponsored at least three *cargos* to rank order 20 of the primary sponsorships in terms of the local *éscalafon*, or graded list of *cargos*. I then calculated the mean rank of these sponsorships in order to provide an informant-provided overview of the hierarchical organization of the *cargos*. I recalculated this after removing each individual's ranking from the total and then correlated his ranking with the composite ranking. This provided a quantitative indicator on how well each individual's *cargo*-holding pattern agreed with the composite ranking of the *cargos* by the other 20 informants (Chick, 1981). Later, Romney, Weller, and Batchelder (1986) proposed cultural consensus analysis, along with a theory of culture as consensus, that provided exactly the same information I had determined, although in a more sophisticated way. Like my 1981 analyses, cultural consensus analysis provides, first, the "culturally correct" answers to a series of questions. In the case of the *cargo* system, these would be questions about the *éscalafon* rank of each of the *cargos*. Second, it provides the degree to which each informant agrees with the culturally correct responses, just as I had done for the *cargo* system in Tepatlaxco. Finally, it provides an overall measure of how well informants agree with the culturally correct answers. I had calculated this, as well, in my 1981 article.

Cultural consensus analysis consists of factor analysis wherein the informants are treated as variables while the variables (the 20 *cargos*, in this instance) are treated as cases. If the factor analysis returns a one-factor solution where the ratio of the first to the second eigenvalue is equal to or greater than three to one and where all of the factor loadings are positive, cultural consensus is said to exist (Romney et al., 1986). Given that this method is superior to my own, in a confirmatory analysis, I used it to determine the "cultural competence" of the 31 informants, that is, the degree to which each agreed with the overall ranking of the 20 *cargos*. Individual cultural competence scores ranged from 0.480 to 0.928 with a mean of 0.805 (SD = 0.098). I then correlated the individual competence score for each individual with the number of sponsorships he had held. As I suspected, cultural competence is unrelated to the number of sponsorships held (Pearson's $r=0.003$, $p=0.99$, $N=31$). This indicates that knowledge of the system is not what differentiates competitors from muddlers.

The above analyses do not demonstrate that participation in the cargo system is, even for some, due to altruism. It is impossible to determine whether the *cargo* system engenders altruism or merely allows it to be exhibited by those who are already altruistic. A safe bet would be the latter. However, I was surprised to find instances where individuals who had already attained elder status in Tepatlaxco (an elder is known by the Nahuatl term *tiaxca*) held additional *cargos*, often at relatively low levels in the system. When I asked about this behavior, several of the elders explained to me that no one was now willing to take on those *cargos* but someone had to do so to keep the system working properly. So, even those who I classified as competitors became cooperators when the system that afforded their status was endangered. Perhaps most important, many of my informants told me that they, and others, participated in the *cargo* system because it provided them with the opportunity to serve their community, despite the often considerable financial cost.

11.2.4 Altruism in the Cargo *System*

Although there was a priest's house in Tepatlaxco and it was, by a considerable margin, the nicest house in the village (I rented it for the summer of 1977), there was no resident priest in the community. The same situation prevailed in nearly all of the villages of similar and smaller size in the area. So, instead of a resident priest, religious functions that required the services of a priest, primarily saying Mass, relied on visiting priests. Visiting priests would travel to villages by car and say Mass or administer other rituals for a fee. For the most part, the villagers despised them because they felt their fees were exorbitant and, moreover, the visiting priests had no real connection to the communities they served. Because of their dislike of the visiting priests, villagers summoned them as infrequently as possible. This meant that as many religious functions as possible fell to the village elders, the members of the *fiscalía*, the lay religious governance body in the community, and the *mayordomos* (Chick, 1981). Together, the *fiscales* and the *mayordomos* composed the *cofradía*, a religious governance institution imported from Spain soon after the conquest.

In addition to its religious components, the *cargo* system provided the range of recreational activities during the festivals, described above. This made the *mayordomos* the primary providers of community-wide recreation in the village (Chick, 1991). Few of these activities had anything to do with religious devotion but they clearly involved community involvement and service. One service was to lure others from nearby communities into the village to spend money or to attract former residents who had migrated, mostly to large cities such as Mexico City, Puebla, or Veracruz, home to visit friends and relatives.

Of the participants in the *cargo* system I studied in Tepatlaxco in 1977–1980, slightly more than half did so for what appeared to be entirely selfless reasons. That is, they got nothing obvious in return for their considerable expenditures in time and, to a lesser extent, money. Before Tepatlaxco had easy access to the outside (prior to 1954, it was connected to Santa Ana Chiautempan only by footpath while, by 1980, several busses per day traveled the 14 miles up and down the mountain between Chiautempan and Tepatlaxco), community members were surely much more interdependent and cooperative behavior was more important than in more recent years. Many of my older informants criticized the lack of community spirit, exemplified by *cargo* sponsorship, exhibited by younger villagers. By 1980, factors such as readily available transportation and work opportunities outside the village appeared to have eroded the system substantially (Chick, 1981, 1989; Dewalt, 1975). Similar systems in larger communities in the region, even by then, had either disappeared completely or been secularized and taken over by local politicians. These degraded systems had religious functions in name only and their real purpose is not community altruism but to bolster local economies, largely through tourism. Few pristine systems remain intact today.

11.3 Conclusion

Thomas Jefferson, in his letter to Thomas Forest in 1814, claimed that nature has implanted in us a love of, and sense of duty to, others. Jefferson seemed to be claiming natural, perhaps even biological, roots for altruism, thus presaging the growing body of research on the place of certain neurotransmitters, such as dopamine, and neuropeptides, including oxytocin and vasopressin, on social behavior. My research in Tepatlaxco did not include research on how neurotransmitters or neuropeptides may have influenced *cargo* festivals in the community. However, these events involved numerous forms of prosocial behaviors such as cooperation, generosity, and positive emotions including anticipation, joy, and excitement. In addition, *cargo* festivals were the most important shared expressive activities in the village and were also the only events that provided community-wide recreation, other than graduation day from the local elementary school (Chick, 1991).

One question, however, is whether the activities of *cargo* holders and other lay religious officers in Tepatlaxco should be considered as altruistic rather than the result of a sense of duty, obligation to their community, or religiously motivated sacrifice or obligation. Clearly, the duties that were part of *cargo* system participation cost sponsors dearly in terms of time and, to a lesser but still important extent, money. Moreover, while some individuals, who I identified above as competitors—those who advanced through the system toward elder status—might have anticipated a reward in terms of status enhancement, the cooperators, who I referred to above as muddlers, had no such anticipation of benefits or gain, at least in this world. Hence, at least for the muddlers, sponsoring *cargos* appears to meet the qualifications for altruism that include being voluntary and intentional and involving concern for the other, empathy, benefits to the receiver, ease of escape, and cost to the initiator, as proposed by Smith, Lapinski, Bresnahan, and Smith (2013) in Chap. 2 of this volume.

It is also possible to think of the *cargo* system in terms of kin selection (e.g., Hamilton, 1964). Members of four village families dominated *cargo* holding and the lay religious system generally in Tepatlaxco although they did not represent a proportionally outsized number of residents. Since present *cargo* system officeholders were responsible for assuring that others would assume their positions for the next cycle, a family in-group bias may have existed, however. There was also a system of fictive kinship, the *compadrazgo* system, at play in Tepatlaxco, as in much of Mesoamerica (e.g., Nutini & White, 1977; Nutini, 1984), that may also have influenced candidacy and recruitment for *cargo* sponsorships (see Chap. 3 of this volume by Coe and Palmer (2013), regarding the place of kinship in traditional societies).

Rituals such as found in the Mesoamerican *cargo* system in Tepatlaxco appear to provide an opportunity for altruistic behavior, particularly in terms of providing recreation to members of one's community. Only when easy transportation and other amenities permitted individual members of communities such as Tepatlaxco to become less dependent on each other did the altruistic aspects of the *cargo* system begin to wane. Other rituals and ceremonies, including various rites of passage where individuals or groups make sacrifices in terms of time, money, or other resources with no anticipation of return, appear to be expressions of altruism, as well.

Acknowledgments This chapter is an adaptation of Chick (2008). Altruism in animal play and human ritual. *World Cultures eJournal, 16*(2). It is republished in this volume with permission.

References

Cancian, F. (1965). *Economics and prestige in a Maya community*. Stanford, CA: Stanford University Press.
Carrasco, P. (1961). The ladder system in Mesoamerica. *American Anthropologist, 63*, 318–334.
Cason, T. N., Saijo, T., Yamato, T., & Yokotani, K. (2004). Non-excludable public goods experiments. *Games and Economic Behavior, 49*, 81–102.
Chick, G. (1980). Concept and behavior in a Tlaxcalan religious officeholding system. Unpublished doctoral dissertation, Department of Anthropology, University of Pittsburgh, Pittsburgh, PA.
Chick, G. (1981). Concept and behavior in a Tlaxcalan *cargo* hierarchy. *Ethnology, 20*, 217–228.
Chick, G. (1989). Expressive and instrumental components of participation in a Tlaxcalan *cargo* hierarchy. In R. Bolton (Ed.), *The content of culture: Constants and variants. Studies in honor of John M. Roberts* (pp. 441–454). New Haven, CT: HRAF Press.
Chick, G. (1991). Acculturation and community recreation in rural Mexico. *Play and Culture, 4*, 185–193.
Chick, G. (2002). Cultural consonance in a Mexican festival system. *Field Methods, 14*, 26–45.
Chick, G. (2004). Rites of passage. In G. Cross, G. Chick, & J. W. Loy (Eds.), *Encyclopedia of recreation and leisure in America* (Vol. 2, pp. 210–218). New York: Charles Scribner's Sons.
Coe, M. K., & Palmer, C. T. (2013). Cross-cultural variation in altruism: Traditional parental manipulation and ancestor–descendant conflict. In D. A. Vakoch (Ed.), *Altruism in cross-cultural perspective*. New York: Springer.

Dewalt, B. R. (1975). Changes in the *cargo* systems of Mesoamerica. *Anthropological Quarterly, 48*, 87–105.
Gibbons, J. L. (2013). Guatemalan adolescents' reports of helping in urban and rural Mayan communities. In D. A. Vakoch (Ed.), *Altruism in cross-cultural perspective*. New York: Springer.
Grönlund, H. (2013). Cultural values and volunteering: A cross-cultural perspective. In D. A. Vakoch (Ed.), *Altruism in cross-cultural perspective*. New York: Springer.
Hamilton, W. D. (1964). The genetical evolution of social behavior. *Journal of Theoretical Biology, 7*, 1–6.
Jefferson, T. (2005, orig. 1814). To Thomas Law Poplar Forest. The letters of Thomas Jefferson: 1743–1826. The moral sense. *From revolution to reconstruction ... and what happened afterwards*. Accessed February 12, 2005, from http://odur.let.rug.nl/~usa/P/tj3/writings/brf/jefl230.htm.
Komorita, S. S., & Parkes, C. D. (1995). Interpersonal relations: Mixed-motive interaction. *Annual Review of Psychology, 46*, 183–207.
Kurzban, R., & Houser, D. (2005). Experiments investigating cooperative types in humans: A complement to evolutionary theory and simulations. *Proceedings of the National Academy of Science, 102*, 1803–1807.
Nash, M. (1958). *Machine age Maya: The industrialization of a Guatemalan community*. New York: Free Press.
Nutini, H. G. (1984). *Ritual kinship: Ideological and structural integration of the compadrazgo system in rural Tlaxcala*. Princeton, NJ: Princeton University Press.
Nutini, H. G., & White, D. R. (1977). Community variations and network structure in the social functions of *compadrazgo* in rural Tlaxcala. *Ethnology, 16*, 353–384.
Romney, K. A., Weller, S. C., & Batchelder, W. H. (1986). Culture as consensus: A theory of culture and informant accuracy. *American Anthropologist, 88*(2), 313–338.
Smith, S. W., Lapinski, M. K., Bresnahan, M. J., & Smith, S. L. (2013). Conceptual aspects of altruism in cross-cultural contexts. In D. A. Vakoch (Ed.), *Altruism in cross-cultural perspective*. New York: Springer.
van Gennep, A. (1960). *The rites of passage* (M. B. Vizedom & G. B. Caffee, Trans.). Chicago: University of Chicago Press. (Original work published 1909).

Chapter 12
Epilogue

To Give or Not to Give: Confessions of a Humanitarian Aid Worker

Lewis Aptekar

"You cry when you go and you cry when you leave."

—Frances Aboud, Canadian psychologist

One fact of humanitarian assistance is that there are always more people in need than resources to help them. If you are going to work in this business, you will have to learn to turn your back. Unless you can get to the point of being able to refuse to administer to a sick child, you will eventually leave, and to leave is to fail.

My first glimpse of this was in the 1980s. I was collegial with an elderly Catholic priest who had been working with street children in Kenya for nearly four decades. If he could be put into any classification of helper, it was not the bureaucrat or the do-gooder. Between the iridescent orange scarf wrapped around his neck that flowed to his knees and the open-sided jeep filled with street boys, he cut a clear, if atypical, picture. Each Monday night he rambled around the central district of Nairobi, looking for street kids to talk to, principally to explain what his programs had to offer should they be willing to join them. It was my pleasure (and learning) to accompany him.

Of the hundreds of *street workers*, which are what the people who go into the streets to help street kids, rather than those who stay in the offices, are called, he was without a doubt one of the best. In fact, at that time in Nairobi, street kids and his name were as invariably linked as apple pie and my country.

On our way home from this particular night, one that I already found noteworthy because of the persistent cold drizzle that I couldn't escape, we had just pulled up to a red light when a girl in early adolescence ran to us waiving her arms frantically. I could see she was not looking well; I thought she might be suffering from malaria. Behind her, in the shadows of a dim streetlight, were six or seven other girls of her age. As she approached the jeep, she pleaded with the priests to take her sick friend to the hospital.

L. Aptekar (✉)
Department of Counselor Education, San Jose State University, San Jose, CA, USA
e-mail: lewis.aptekar@sjsu.edu

D.A. Vakoch (ed.), *Altruism in Cross-Cultural Perspective*,
International and Cultural Psychology, DOI 10.1007/978-1-4614-6952-0_12,

Without much of an excuse, he refused. He told her friends to take her to the hospital. All he offered was to check on her in the morning. As we sped off toward where I was staying, I was confounded. Why had he left the girl in such a crisis? He told me that it was past ten o'clock at night, and if he took her to the hospital, he wouldn't get home until past one in the morning. He had mass to give at six and a full day of street work already planned to do afterward: "I have to draw the line somewhere."

I felt betrayed. He wasn't living up to the standards of his religious calling, let alone the secular morality that is associated with humanitarian assistance. It took me several years to understand the priest's refusal to administer to this sick girl. I had to suffer more than a few doses of self-incrimination before I got to a place where I could draw my own line in the sand. What follows is my attempt to convey how I live with myself in situations where there is always less time and energy than work to be done and where the results of failing to meet the demands can be, and often are, significant.

In 1996, I went to Ethiopia on a 2-year contract to study the mental health problems of people displaced by the civil war. Most of my work was in a single camp, Kaliti, which was on the southern end of Addis Ababa, the capital. Kaliti on one dusty acre had a population density greater than that of the Warsaw Ghetto. There was no source of water, a single latrine, and no electricity. The people in the camp were allowed to work in the public domain, earning 3 kg of grain per day, per family, but because the average family size was nearly eight and because the grain they earned did not always arrive on time, they lived with little food and the constant fear of having none. When I arrived, they had already lived like this for 6 years.

Before they arrived in Kaliti, the men had made a decent living as civil servants; however, when Eritrea won the civil war in 1991, their wives—mostly women from the province of Tigray now in the new state of Eritrea—were given only a few hours to choose either Eritrean or Ethiopian citizenship. If they opted for Eritrean citizenship, they would be close to their families of origin, but they would lose their husbands and children, who were considered enemies of Eritrea. If the women accepted Ethiopian citizenship, they could continue living with their husbands and children but had to leave their families of origin, and to get to Addis, they were forced to trek through the Danakil Depression, arguably the most inhospitable spot on earth. There is no water, and the temperature can reach 50 °C in the shade (122 °F). Because they had been swept hurriedly from their homes, they left with only what they could carry in their hands and on their back and heads. Insufficient water became the rule, and before they arrived in Ethiopia, almost all of them bore witness to relatives and friends who had died from thirst.

In Kaliti, acute conditions were common, the depth of which was rarely encountered in the West. Typhus, typhoid, and malaria were common. TB and AIDS were endemic. I must admit that being exposed to this had its attraction. I was not there to be altruistic but to advance myself. What I learned came from finding myself a wealthy philanthropist. I was forced to take on a common man's dream, "what if you won the lottery ticket and with it millions of dollars, how would you use it for the betterment of humanity?" I found this an immense burden and did my best to avoid taking it on, but correspondingly, I found an opportunity to come to terms with a part of myself rarely engaged in my work as a professor at San Jose State University, where what is expressed as unconditionally necessary is beyond comprehension to the absolute needs of the people in Kaliti.

On one of my first visits to Kaliti (and what was to become characteristic of my work for the next 2 years), I was taken to help Aster, a small dark-skinned woman in her mid-30s. She too was so acutely ill that she could not raise her head to acknowledge me when I entered her cardboard tent to take her history. She had a high fever and was dehydrated. Her friend Checkla told us that Aster had not eaten for several days. She said, Frazier, Aster's son, who was her only living relative, sat by her side, remaining with her night and day, bearing witness to her demise.

Among humanitarian workers in Ethiopia (and most probably in most places), there was a commonly debated question (and one that paralleled the welfare debate in the West): Should *material* aid be given to the poorest of the poor? Those who were against this type of help argued for teaching fishing instead of handing out the fish. They talked about "aid dependency," an argument common in developed countries as well.

My reasons for helping materially boiled down to the following factors: (1) by allowing us to conduct research in the camp, Kaliti people earned it; (2) by inculcating a therapeutic trusting relationship in our exchanges, our research and community counseling were facilitated; and (3) because we had it, they needed it, and no ethical alternative existed.

What was missing from the humanitarian debate in Ethiopia was the other side of the coin of aid dependency, aid fatigue. This condition focuses on the difficulty of helping and is far more complex and important to consider. First, as deep as my purse was, it was not sufficient to take care of everyone who needed help. Second, every attempt to help was diverted into a promise for full care, making it nearly impossible for me to get satisfaction for what I did give. And, finally, whatever I gave resulted, because of the endless need, in a lack of progress. Nothing is more debilitating than altruism without satisfaction.

Aster's immediate concern was that she didn't have enough food with which to take her medicine; each time she tried to swallow the medicine that would make her better, she vomited. Without food, she couldn't keep the medicine down. With a grunting physical effort that was more eyes than words, she pleaded with me for money for injections; injections can be taken without food.

I found that it is far more difficult to ignore the needs of a single known person in crisis than to refuse the fate of unknown traumatized multitudes. In this way, I was similar to how I observed expatriates refusing to help beggars in the streets of Addis. In spite of the opportunity to help the beggars, people didn't, because to give was to acknowledge the person doing the begging, while refraining allowed a way to avoid such dismal reality.

While it might seem surprising in the face of seeing someone so much in need, before it was possible to help that person, it was necessary to acknowledge one's own suffering. In the face of someone truly in need, like Aster, we are made uncomfortable by our powerful need to flee. Only by accepting that we too are in need can we stand face to face in the presence of someone else's begging for help.

The truth for me was that I had no choice; I simply was not comfortable being so close to people in need and being able to help and not doing so. My struggle was more on the other end: Did I give enough? Why did I not give more? What kind of person was I? Maimonides said, only a man with an evil eye would give less than 10 % of his wealth. I was not even close to this.

Faced with myself in this type of situation on a fairly daily basis, I found myself preoccupied with the diverting idea that the people of Kaliti saw me only as someone who gave them money or food or medicine. I thought they were not getting the best of what I had to offer. I had hoped to give from the base of my professional knowledge and any wisdom I had garnered from my experience in working in diverse cultures with individuals coping in difficult circumstances. After all was said and done, whatever the degree of destitution Kaliti people might have to face, I wanted them to understand that any basis for separating the resilient from the desperate had *more* to do with the human spirit than any material benefits coming from me.

But Aster, who had already spent all her resources on medicine, which, in fact, was bought with the money that was supposed to have been spent on food, had more immediate concerns. By my figuring, there were three ways for her to get money: (1) She could borrow it from a family or friend. (2) She could borrow it against her future grain rations. (3) She could get money from me. Her first option was no option. Her family had all perished except for her son, Frazier, who was already contributing as much as he could by doing child labor in an economy that had no jobs for anyone. And her neighbors already loaned her some money; they could give her more but only by jeopardizing their own lives. Her second option was also not viable because she had already borrowed against her future grain rations once. If she did this again, she would have no hope of being able to retire her debt and, upon her death, the debt might be passed onto her son. This left the third option: me.

When I put my hand in my pocket, I could feel enough money to pay for the food she needed and, for that matter, enough to keep death at bay for many people who I would come to see nearly every day for these 2 years—people whom I would come to call friends. Another diverting strategy I took, before accepting what I had and they didn't, was to watch myself through their eyes. They also had to deal with giving or not giving, in some instances having to choose between feeding one's own child and letting the other sicker one pass away.

As I stood there in front of Aster, I told one of my colleagues we would have to take money out of our budget to start a fund for this kind of emergency. Then I walked on to continue my rounds thinking that if I take care of Aster at the beginning of the day, then where would I find the resources for the next Asters that I was sure to see before the day closed? I couldn't blame them for their demands, yet I found I couldn't stand up to helping them.

One way I made it through the day was keep the people in Kaliti shrouded in abstractions of misery. The truth is that real tangible misery is something that affects me personally as it did the few times someone in my family was in trouble, like when my wife's malaria was on the verge of taking her over. These people were objects of study, allowing me a "professional," "objective" view.

In this context, I asked: Did the people in Kaliti grow accustomed to suffering? The answers from my Ethiopian colleagues were not conclusive. But I did learn that parents believe that children have their own God who takes them when the time is right. Parents leave a single tuff of hair on their children's otherwise shaved heads so that God can easily take them. Do these beliefs make it easier to lose your child?

A few weeks later, Aster was admitted to Mother Teresa's home for AIDS victims. Aster's friend Checkla, who took her there, was staying with her because Aster needed constant care. Patients were not allowed to enter Mother Teresa's home without providing for their care. Unfortunately, a problem emerged—since Checkla would be away from the camp caring for Aster, she would not be able to work for food, and who would take care of feeding Checkla's two children?

To keep myself afloat in this work, I was going to have to see how far the following questions apply to me: Am I my brother's keeper? Were the people in Kaliti foreign in nearly every aspect of my life, sharing only the abstraction of humanity? Could I learn to operate from the stance that they did share full brotherhood? Or, would I draw a line of continuum—much for my blood brother, less for my extended family member, and so on?

After Aster spent a week or so at Mother Teresa's, the level of conversation about who was going to take care of Checkla's children escalated. At one point, the conversation became heated. Amharich, a flamboyant, overly charged, stout middle-aged woman, screamed at me to get out of the camp. We were not helping them, she shouted, only causing more problems. Why should she give up her grain ration for Checkla's two children when I could give them money?

I am thinking that if I support Checkla's children when would I stop? Would I wait for Aster to die before I stop? Or should I save my resources for Checkla's kids? If I didn't help, at what point would the community stop helping because they think Aster was going die with or without help? Was this argument they were having their way of talking about this, of dealing with the dilemma of not having enough to live up to their own standards of generosity?

The conversation about how to feed Checkla's two kids reached such ferocity that I announced I was leaving to visit Aster. I take along with me a couple of colleagues from Kaliti. On the way, one of them tells me that Checkla's motivation for helping Aster is financial. She is behaving altruistically by betting that we will give something to her kids if she is not around to feed them.

When I get to Mother Teresa's, I discover that Aster died that morning. Back at camp, the conversation streams in two directions. One is Fraizer, her surviving 8-year-old boy. Another one of Aster's friends, Lumlum, agrees to take care of Frazier until something more permanent can be worked out. What will I do with helping Lumlum take care of Frazier? Before I can deal with this, an argument commences about what to with Aster's body. Some argue that it should be left at Mother Teresa's, which would mean that Aster would not receive a proper burial. This position is based on the costs of bringing the body back to Kaliti. Would it not be better to use all the resources that Aster had (she had some 100 pounds of *teff*, the ancient wheat species endemic to Ethiopia, which is the dietary staple) for taking care of Frazier, now an orphan? Amharich wants Aster's body returned to Kaliti and prepared for a proper funeral. She claims that since Aster has been in the camp with the rest of them for 6 years, she should be buried among them.

Often, helping has less to do with whether or not I was going to give than with the logistics. Behailu, the chairman of the elected camp committee, steps in and says Aster deserves a proper burial. The conversation stops and we drive into Mother Teresa's to get the body. On the way, I decide to pay Lumlum for taking care of Frazier even though I knew she might skim a bit off to feed herself. I also thought she might find a way to do the same for what she got from selling Aster's *teff* that was originally supposed to be used for the burial and for the priest who will bury her.

When we get to Mother Teresa's, we discover—without going into the details of how we come to learn this—that the body doesn't fit into the car. We need another one. If I use the money for transport, there will be nothing left for the burial. And so it goes. Each time I give, I know that one takes the money and spends it in a way that does him benefit. The other complains of not being considered equally. If I chose to give to my friend, after all what are friends for, the elected committee complains, but those in charge are more selfish politicians than trustworthy administrators.

Sometime during this imbroglio, I visited Sister Mary, a middle-aged Italian nun with a sharp smile and an oversized crooked nose, who had spent some 20 years feeding the indigent in Addis Ababa. I was interested in how she had managed the logistics. Each Wednesday, mothers would bring their sickly children to her, and she would attach color-coded flags to their wrists that ranked just how sickly they were. She strictly constrained her feeding program to fifty children and their mothers, despite pressures to take on more. I asked how she dealt with the pressures. She told me that I must realize "the desire to help could never be totally consummated." There was only so much she (or anyone) could do, and it was best to define (and limit) one's offerings.

In fact it is not divisible. The most important Ethiopian Orthodox legal document, the *Fetha Nagast* (*Law of Kings*), written in the seventeenth century and still used for ecclesiastical law today, stipulated several rules pertaining to the responsibility of giving to the poor. If a person gave to the poor, that donor earned the support of God. The more one gave, the more likely he or she would gain God's favorable attention. God not only gives a place in the kingdom of heaven to everyone who gives to the poor but also the poor provide an opportunity for the wealthy to grow closer to God. This is a two-way street.

When I was leaving Sister Mary's compound, a young Oromo woman with a small child stood at the gate, waiting to see the Sister. The little girl had been burned over her face, body, and arms. Wound around the child's head was a soiled piece of white cotton cloth, wrapped as if she had a toothache. An alarming red pimple protruded from the cornea of one eye. In reaching out to touch the Sister, I could see the little girl's fingers, burned to stubs. Yet, Sister Mary welcomed the child in her arms, neglecting the striking blemishes, and admonished the mother only about the girl's runny nose. To keep at this work, she had found a way to ignore the obvious and enjoy the unexpected, and she had developed a philosophy based on her own caveats, personal judgments, allowing for plenty of exceptions, and more than an occasional breaking of rules.

Well, Aster's body is brought home. She is buried properly. Checkla returns to her two kids. Frazier goes to live with Lumlum. We look for an orphanage for him but get bogged down in bureaucracy. Lumlum, a widow without children, doesn't want him to leave. Several months later, she thanks me for allowing her to take care of Frazier. I am ecstatic that my original fears, that Lumlum only offered to care of Frazier to feed herself, had some altruistic motive.

As I left the camp for the last time, my friend's friend, whose leg was broken in a soccer game, asks me to help him with the medical care that will allow him to walk again. When I try to explain to him that I have nothing for him, he asks, although clearly it is meant as rhetorical, "What kind of a humanitarian aid worker are you?"

Now, several years later, I would like to say to him: I accept that that instead of being applauded for what I did, I could be harassed for not doing enough, but feeling harassed stopped me from giving any more than I did. I admit that to the extent that I took care of "others," I was also taking care of my needs to be a worthwhile person, but there was more to it than that. It was a two-way street. I also helped because it gave me a chance to live deeply, perhaps being religious, perhaps living in God's name. What I tried to teach them was to increase their level of citizenship and found instead that they taught me (like Sister Mary) to reduce mine. Likewise, they taught me the value of spirituality, while I tried to impress upon them to become less dependent on God's will. In the end, each of us decided what, if anything, there was to learn from the other.

What all this self-exploration revealed to me about giving—and what I learned in talking with people who had discovered how to sustain inner resolve in helping (as well as to those who had dropped out)—proved to be elegantly simple: The only way to continue giving is to drop any pretense of impartiality, even fairness, and just accept one's desire to help whichever people one liked for whatever reason. The formula may not be politically correct, but it works, and in the end, the needy will fare the best from its application.

I want to tell humanitarian workers not to withdraw from the overwhelming need, even knowing that resources will be inadequate to solve all the problems and that the demands are never ending and often self-serving. Find a way to enjoy what you can offer, and ignore the rational reasons for not giving. Also, don't assume that the degree of giving is stable over time or place. It appears to be as much of a state as a trait. My openness to giving is not the same in America as it was in Ethiopia. In fact my experiences in Ethiopia have led me to see American beggars here in the context of Ethiopia, which doesn't do much for them.

In Kaliti, the good and the bad coexisted. Illness and disease provided the awful drama of slow and tortured death as well as the opportunity for love and caregiving. In observing the many acts of kindness characterizing the relationships around me there, I came to understand how important the very act of caring can be.

Chapter 13
Afterword

Stephen G. Post and Matthew T. Lee

In the most general and accepted terms, *altruism* is any behavior or attitude that can be contrasted with *egoism*. For the altruist, others are "ends-in-themselves" rather than mere means to the fulfillment of the interests of the self, which is what defines the egoist. *Altruism* means, literally, "other-regarding." As such, it would seem to be a plausibly universal attitude and behavior, at least with respect to the nearest and dearest, such as family members and friends. The extension of altruism to wider circles is commonplace within the narrow context of the agent's communities; extension to a shared humanity may sometimes be limited by intense group loyalties but is also clearly achievable through helping activities that transcend ethnic, cultural, religious, racial, class, and species barriers. At least in principle, the universality of altruism is best demonstrated by the universality of the Golden Rule in its form of "doing unto others," which appears widely across cultures and times, and is thus an aspect of what Aldous Huxley termed *the perennial philosophy*. In broad terms, other-regarding behaviors are without exception endorsed in all major world religions and in the world cultures that have grown up around them.

Does altruism require self-sacrifice? Not in any essential way. Often the altruist identifies with the needs of the other, and therefore, the sense of cost more or less vanishes, unless the activities are unduly strenuous and exhausting. The altruist may engage in face-to-face helping, contributing, and serving others in areas of legitimate need without experiencing this in any psychological sense as a burden or as costly to self. This is the case even when an outside observer might be able identify significant costs, such as when a medical volunteer contracts a debilitating disease while meeting the health needs of others and shrugs it off as an expected

S.G. Post (✉)
Center for Medical Humanities, Compassionate Care, and Bioethics,
Stony Brook University, Stony Brook, NY, USA
e-mail: stephen.post@stonybrook.edu

M.T. Lee
Department of Sociology, University of Akron, Akron, OH, USA
e-mail: mlee2@uakron.edu

D.A. Vakoch (ed.), *Altruism in Cross-Cultural Perspective*,
International and Cultural Psychology, DOI 10.1007/978-1-4614-6952-0_13,

occupational hazard. Such altruists have redefined costs and benefits in ways that may puzzle an observer. Indeed, in many instances, such behavior allows the agent to feel deeper purpose, gratification, and happiness. Altruism therefore does not include self-sacrifice in its core definition, unless one defines such self-sacrifice in terms of any activity that is not directly focused on the narrow interests of the self. But this makes no sense given the essentially social and communal way in which humans flourish. In general, the interests of the self and the interests of others whose security matters to the agent are coincident or coextensive, although this sense of commonality may dissipate under conditions of resource stress.

There are times, then, when caring for the other really does cut into the well-being of the agent, and in the absence of community support or respite, this can take a toll emotionally and physically. For many people, there are spiritual sources of empowerment and meaning that can mitigate the mental and physical costs, on balance. But it is certainly possible, although not required, that a sense of disadvantageous self-sacrifice can be associated with altruism, in which case altruistic duties may become more central to moral consciousness than do altruistic dispositions. Comte, the sociologist who coined the term *altruism*, understood loss to the agent as a possibility, but not a necessity for behavior to be deemed altruistic. It is only in modern biological circles that loss to self has been deemed an essential aspect of altruistic action. Their paradigm is the animal that makes a warning cry to protect others from an attacker. But nonhuman animals do a great deal of helping behavior that is routine and unassociated with risk. As with humans, risk only sometimes comes with the territory of altruism. One wonders why those who study animal behavior focus on warning practices, rather than on more general altruistic behaviors.

It is essential that altruism, to be such, must be separated motivationally from the interest in reciprocal gains. These gains may come, and hopefully will, but they are a secondary motivation rather than a primary one. Reciprocal altruism is less altruism than contract. We do believe that altruism, to be genuine, must primarily involve the motivation to help others, not self, although internal benefits to self may and usually occur regardless. If reciprocal gains follow, they are welcome, but not sought as a goal. But once again the spiritual dimension complicates the picture somewhat: altruists may be motivated to follow a divine calling, which often includes helping others. The primary motivation here may be to serve God and do God's will, with secondary consideration given to benefits that may accrue to self or others. Habits of helping can become internalized and almost a matter of reflex, so that the "motivation" may involve the inclination to know God's love. Expressing this love to others has been experienced in some cases as basically a conditioned response.

In this collection, certain chapters do a splendid job of contextualizing altruism within major world traditions, and they seem to agree that altruism, volunteering, service, and helping others are evident in these various traditions. These chapters add specificity to the religious cultures of altruism. For example, Joan D. Koss-Chioino finds that altruism is integral to both spiritual transformation and healing practices among Spiritists in Puerto Rico and the United States. Contrary to the egocentric understandings of self that are prevalent in modernist societies, premodern groups are portrayed as advancing "sociocentric" worldviews that foster deeper levels of empathy. Spiritists, for example, practice a "radical" empathy that extends not only to the individual who has sought their assistance with healing but also to all others who may need healing. This deep and extensive empathy may cause the healer to suffer greatly. But to further complicate matters, according to their belief system, the ongoing health and well-being of the Spiritist healers themselves depends on their willingness to engage in precisely this kind of other-directed helping behavior. Suffering for the other is required to heal both self and other. This is why Koss-Chioino concludes that "ritual healers feel they have no choice but to carry out the healing work."

Yueh-Ting Lee and colleagues from the University of Toledo, in "Daoism and Altruism: A China-USA Perspective," write about a very different religious and cultural context in their exploration of the relationship between altruism and ancient Chinese Daoism (or Taoism). It is fascinating to learn about "wateristic" personally attributes. Altruists should be like water, "modest and humble," always going to "the lowest place," and "helpful and beneficial to all things." Water is "flexible" and "transparent" and exhibits "gentleness with perseverance." Lee and colleagues tell us that water is altruistic because everything depends on it, but it seeks nothing in return. It goes to the lowest level, like a humble person who does not wish to be aggressive or competitive. Water is soft and gentle, but over time in persistence, it will cut through the hardest rock. The Daoist goal is to cultivate a water-like personality. There is a most interesting cross-cultural affinity between this Daoist ideal and some of the Christians engaged in benevolent service that we report in our coauthored book with Margaret Poloma, *The Heart of Religion*. Like the Daoists, some Christians argue that the true apostle of God is the one who "goes to the lowest place" to serve and empower others in need. The servant of God does not seek self-aggrandizement by keeping the helped in a permanent state of dependence, forever subordinate to and reliant on the helper. Instead, the goal is to lift up the needy so that they are not only able to meet their own needs but become benevolent leaders who in turn help others in need along this path. And like Koss-Chioino's Spiritists, some of these Christian altruists also feel that they have no choice but to practice radical and extensive empathy. There do seem to be important commonalities across cultures in this regard, despite the coexistence of differences.

Other chapters engage with this important point in nuanced ways. For example, Abhik Gupta, in "Altruism in Indian Religions: Embracing the Biosphere," identifies a number of commonalities across diverse Indian religious traditions. The ideal of universal "non-harm" and "biosphere altruism" is facilitated by diverse cultural forces ranging from tribal creation myths involving a "primordial altruism" to the centrality in Buddhism of *Metta* ("loving kindness") meditations. Alexandra Arkhipova and Artem Kozmin note, in their chapter on cross-cultural altruism in folktales, that in Germanic and Baltic cultures, there is often an emphasis on saving the self by one's own efforts. However, this does not diminish the theme of helping others, and most of us can recount numerous fairy tales in which beneficence wins the day. Such themes seem more prominent in Eastern European folktale traditions. Bruno Bettelheim, for example, in his classic *The Uses of Enchantment*, shows how often fairy tales establish the values of kindness and helping in the minds of children. In her chapter "Cultural Values and Volunteering: A Cross-Cultural Perspective," Henrietta Grönlund indicates that volunteerism is essentially universal, but shaped by cultural factors. She also points out that cross-cultural variations in volunteerism are still in their infancy.

The question of extending altruism to a shared or a common humanity is really the key interface, for at the intersection of groups of various kinds, we sometimes get the impression that altruism is anything but universal and is easily overwhelmed by harmful tendencies and behaviors. Altruism can be extended to a shared humanity by appreciative understanding of the traditions of the other, by a moral principle of equal regard, by an abiding compassion regardless of differences, by role training or positional expectation as with helping occupations, and by spiritual experiences of an underlying reality of love in the universe. Actually, when we see acts of violence and hatred, we are usually witnessing either narrowly focused altruism and love that has assumed a defensive morality vis-à-vis outsiders or destructive emotional states overwhelming altruism for various reasons. But the altruism is there, even if difficult to discern.

Spiritual and historical cultures are vital to the extension of altruism to a universal concern. In medicine, as an example, ethics rightly begin with the writings of Hippocrates (400–300 BCE). Yet one finds in the ancient Greeks and Romans no passionate or compassionate concern for the patient. Unlike the Good Samaritan, the Hippocratic physician does not go out of his way in response to patient needs. In fact, barbarians, slaves, poor people, the contagious, and dying patients were to be ignored and certainly were not part of the physician's domain of duty. While the great Hippocratic tradition has great strengths, it has these deep limitations as well. It is only the Hippocratic Oath as it later absorbed the light of universal and unconditional love of the patient in the great period of the Judeo, Christian, and Islamic growth that we hold dearest. Here the physician is no longer casual but rather called by God to heal the sick regardless of their circumstances, degree of illness, or ability to pay. The Islamic Code of Medical Ethics as it is articulated today reflects this depth of calling to serve the needy. The physician swears to Allah to protect human life in all stages and in any situation, doing his or her "utmost to rescue it from death, malady, pain and anxiety." The physician protects dignity and is "an instrument of God's mercy, extending my medical care to near and far, virtuous and sinner, and friend and enemy." One finds a similar depth of commitment in Moses Maimonides, whose famous prayer hangs on the wall of countless Jewish clinicians. It reads, "The eternal providence has appointed me to watch over the life and health of Thy creatures." Furthermore, "May the love for my art actuate me at all times; may neither avarice nor miserliness, nor thirst for glory of for a great reputation engage my mind; for the enemies of truth and philanthropy could easily deceive me and make me forgetful of my lofty aim of doing good to Thy children." These lines are entirely different in tone and passion for the needy patient than anything that could possibly have been produced by the Hippocratic Oath. The oath makes reference to the Greek gods and goddesses, but it has no such depth. From the Prayer of Maimonides to the Christian founding of the first hospitals, from the advances made by Muslim physicians to the establishment of great medical schools in Europe and the Middle East, from Florence Nightingale's founding of modern nursing to Dame Cicely Saunders' establishment of the Hospice movement, and from Albert Schweitzer's "reverence of life" to Paul Farmer's "theology of liberation," good medical practice owes so much to a sometimes forgotten God of love. Those who have no interest in God can still appreciate your being committed to walking with them through their experiences, on their terms, and this is more than sufficient testimony to the power of love in all of our lives.

Is altruism a universal value? It would seem so. Helping others is a necessity in any community. As Darwin pointed out in *The Descent of Man*, much of human evolution occurs between communities such that those communities with the highest degrees of helping, serving, caring, and altruism are likely to survive and prosper. However, achieving loyalty to the universal community remains the challenge of our times. Most people of good will believe that we will get to such universality because the survival of the human species depends on it. It is possible that an unlimited, altruistic love may be moving toward the status of master imperative in a number of cultural traditions. Such love would serve as a lens through which the religious texts of these cultures are read and reinterpreted. In this sense, religion could increasingly serve as an additional facilitator of the universality that evolution has helped to instill in all of us, regardless of cultural background. It goes without saying that both evolution and religion have also fostered insularity and intergroup conflict. The chapters in this book leave us cautiously optimistic that even more common ground across cultures can be identified with regard to altruism.

This volume is an initial exploration of a theme that requires more work. It is clear that the "love of neighbor as self" is virtually a universally stated precept, and we await the time when its implementation will be equally universal.

About the Contributors

Lewis Aptekar, Ph.D., is Professor of Counselor Education at San Jose State University. He received his doctorate in educational psychology from the University of Michigan, and he completed his postgraduate training in clinical psychology at Texas Tech Medical School. Dr. Aptekar is a past president of the Society for Cross-Cultural Research, and he is licensed as a clinical psychologist. His academic awards include two Fulbright scholarships (Colombia and Swaziland), a Senior Fulbright Scholar award (Honduras), Nehru Visiting Professor (University of Baroda, India), a Kellogg Foundation/Partners of the Americas Fellowship in International Development, a Rotarian International Ambassadorship (Zambia), and a scholarly residency at the Bellagio Rockefeller Foundation Study and Conference Center (Italy). Dr. Aptekar's books include *Street Children of Cali*, *Environmental Disasters in Global Perspective*, and *In the Lion's Mouth: Hope and Heartbreak in Humanitarian Assistance*.

Alexandra Arkhipova, Ph.D., Humboldt Fellow, is Associate Professor in the Centre for Typological and Semiotic Folklore Studies at the Russian State University for the Humanities (Moscow). As a specialist in contemporary folklore and mythology, her research on Soviet humor led to the publication of the books *Jokes about Stalin: Texts, Comments, Research* (with Mikhail Melnichenko) and *Jokes and Movies*. Dr. Arkhipova's recent research has focused on examining Mongolian and South Siberian mythologies in a cross-cultural perspective. She is currently working on the project "Anthropology of Money," and she is the coeditor (with Jacob Fructhman) of the book *Anthropology of Money in Post-Soviet Russia*.

Mary J. Bresnahan, Ph.D., is Professor in the Department of Communication at Michigan State University in East Lansing. Her interests in health communication include cultural factors that impede organ donation in East Asia, adolescent smoking prevention messages in China, and stigma and body size. Dr. Bresnahan also studies cultural differences in corrective facework after a face-loss event.

D.A. Vakoch (ed.), *Altruism in Cross-Cultural Perspective*,
International and Cultural Psychology, DOI 10.1007/978-1-4614-6952-0,

Sydney Xinni Chan, M.A., is a doctoral student in experimental psychology in the University of Toledo. She received her B.A. in psychology from the University of Kansas in 2006, with an honors thesis on interpersonal relationships between East Asians and Westerners under the guidance of Dr. Glenn Adams. She obtained her M.A. in industrial organizational psychology from Wayne State University in 2008. Her research interests include stereotype (in)accuracy, cross-cultural social perception, and intergroup relations. In particular, she has been working on the Evaluation-Potency-Accuracy theory of stereotypes and its application to cross-cultural stereotypic perception with Dr. Yueh-Ting Lee since 2010. She is a member of the Mid-Western Psychology Association and the Society of Personality and Social Psychology.

Wenting Chen, M.A.Ed., received her B.A. in English literature, and she also holds a master's degree in English as a second language (ESL). She is currently a doctoral student in curriculum and instruction in education at the University of Toledo. Her professional interests include curriculum design for ESL learners, educational psychology, and multicultural education. She has served as a tutor for ESL coursework at the University of Toledo. Her conference presentations include papers presented at the Ethnographic and Qualitative Research Conference and at the Eco-Justice Conference.

Garry Chick, Ph.D., is Professor and Head of the Department of Recreation, Park and Tourism Management at the Pennsylvania State University. He received his doctorate in anthropology from the University of Pittsburgh in 1980, and his primary research interests are in expressive culture, including leisure and health, adult play and games, and research methods. Dr. Chick has conducted field research in Mexico, the northeastern United States, mainland China, and Taiwan. He is a past president of the Association for the Study of Play and the Society for Cross-Cultural Research. In addition to Penn State, he has taught at the University of Illinois at Urbana-Champaign, where he received fellowships to the Center for Advanced Study and the Center for the Study of Ethics and Values, the University of Wageningen in the Netherlands, and the University of Costa Rica. He was the founding editor of the journal *Play & Culture* and is a past editor of *Leisure Sciences*.

Kathryn Coe, Ph.D., is Professor and Lilly Scholar in the Department of Public Health, College of Medicine, Indiana University-Purdue University, Indianapolis. Her doctoral degree is in cultural anthropology and evolutionary biology. Her research interests are broad and include studies of the transition between trial law and law of the state, the evolution of moral systems, the effect of deforestation on the health and well-being of a rainforest tribe in Ecuador, nausea and vomiting during the pregnancies of urban Ecuadorian women, and both infectious and chronic diseases. The specific aspects of culture that interest Dr. Coe are those that are not often a focus in public health, such as the arts, religion, systems of behavioral codes, educational practices, and kinship systems. Methodologically, she favors triangulation of methods—qualitative and quantitative methods combined and community-based participatory research. She has published over 80 peer-reviewed papers on the theme of the evolution of culture and modern humans.

Juris G. Draguns, Ph.D., D.H.C., is Professor Emeritus of Psychology at the Pennsylvania State University. He holds a doctorate in clinical psychology from the University of Rochester. Throughout his career, he has been active in cross-cultural and international psychology, with a focus on psychopathology, personality, and intercultural and interethnic relations. He has taught and lectured, in six languages, in Australia, Estonia, Germany, Kuwait, Latvia, Mexico, Sweden, Switzerland, and Taiwan. Dr. Draguns is author or coauthor of over 170 articles, chapters, and monographs in psychological, psychiatric, and anthropological publications, and he has coedited 16 books, including 6 editions of *Counseling Across Cultures.* He has served as president of the Society for Cross-Cultural Psychology, he was awarded an honorary doctorate from the University of Latvia, and he is a recipient of the American Psychological Association Award for International Advancement of Psychology and of the Penn State Emeritus Distinction Award.

Judith L. Gibbons, Ph.D., is Professor of Psychology at Saint Louis University and Visiting Fulbright Scholar at the Universidad del Valle de Guatemala. She has published widely on adolescence in an international perspective and intercountry adoption. With Karen Smith Rotabi, she is the editor of *Intercountry Adoption: Policies, Practices, Outcomes*. In 2011, she won the Interamerican Psychologist Award for contributions to psychology in English, and in 2012 she was honored with the Florence L. Denmark/Mary E. Reuder Award for Outstanding International Contribution to the Psychology of Women and Gender. Dr. Gibbons is a past president of the Society for Cross-Cultural Research, the editor of the American Psychological Association's journal *International Perspectives in Psychology: Research, Consultation, Practice*, and the incoming president of the Interamerican Society of Psychology (better known by its acronym in Spanish, SIP).

Henrietta Grönlund, D.Th., is a Postdoctoral Researcher in the Department of Practical Theology at the University of Helsinki. In this position, she conducts research on helping, both individual forms (volunteering, giving) and institutional forms (welfare, Church diaconia, nonprofits). Her main areas of interest are the societal positions and roles of religions and churches and the role of values and meaning systems in pro-social behavior. She has studied themes related to volunteering among young age-groups, students, and the elderly in several Finnish and international research projects. Along with several monographs in Finnish, her articles have appeared in *Voluntas: International Journal of Voluntary and Nonprofit Organizations*, *Journal of Church and State*, *Journal of Academic Ethics*, *Journal of Non-profit and Public Sector Marketing*, and several Finnish journals. Dr. Grönlund is currently also Director of Civic Engagement in a Finnish social sector nonprofit called HelsinkiMissio.

Abhik Gupta, Ph.D., is Professor in the Department of Ecology and Environmental Science at Assam University, which is located at Silchar in Assam, India. As a zoologist and animal ecologist, he started his research by studying the ecology of hill streams in the mountains of Meghalaya, India, and later on the bioaccumulation and toxicity of heavy metals in aquatic ecosystems. Presently, he is also investigating with his students the issue of arsenic contamination of groundwater in Northeast

India, which constitutes an important concern for human and ecosystem health. Dr. Gupta has also written on the intricacies of human altruism towards nonhuman plants, animals, and even nonliving entities as well as the role of nature religions in generating and nurturing this form of altruism in India. He is also interested in the ethical, ecocentric facets of traditional community conservation through sacred groves and other mechanisms.

Joan Koss-Chioino, Ph.D., is Professor Emerita of Anthropology at Arizona State University and Research Professor of Psychology at George Washington University. As a psychological anthropologist, she works at the interface between anthropology, psychiatry, and psychology. Her primary interest is the etiology and treatment of illness and emotional problems, whether conventional, alternative, or psychotherapeutic. She has worked in Latino cultures in the Southwest (treatment research with Mexican and Mexican American youths and families), in Puerto Rico for four decades (spirituality, distress, and mental illness among healers and sufferers), and Spain (depressed women). Dr. Koss-Chioino's books include *Women as Healers, Women as Patients: Mental Health Care and Traditional Healing in Puerto Rico; Medical Pluralism in the Andes; Working With Culture: Psychotherapeutic Interventions with Ethnic Minority Children and Adolescents* (with Luis A. Vargas); *Working With Latino Youth: Culture, Context and Development* (with Luis A. Vargas); and *Spiritual Transformation and Healing* (with Philip Hefner).

Artem Kozmin, Ph.D., was a Research Fellow at the Russian State University for the Humanities. On February 1, 2013, he died at the age of 36 in Ulan Bator, Mongolia. In 2003, he completed his doctorate in folkloristics on tale-type and motif indices. He conducted fieldwork in North Caucasus and Mongolia, and he worked on a digital database containing folklore texts in Polynesian languages. His research interests included the areal distribution of fairytales and folklore motifs, formal approaches in text analysis, and non-European folklore. Dr. Kozmin also participated in projects on political anthropology. He was the author of *Plots of Fairy Tales: System and Structure*, which explores the various formal features of fairytale plots. He received grants for his research from the Ford Foundation, INTAS, Deutscher Akademischer Austausch Dienst, and the Russian Foundation for Basic Research.

Maria Knight Lapinski, Ph.D., is jointly appointed as Associate Professor in the Department of Communication and Michigan AgBio Research at Michigan State University (MSU). She also serves as the Associate Dean for Research at the College of Communication Arts and Sciences. She received her M.A. from the University of Hawaii, Manoa, and her doctorate from MSU. Her research examines the impact of messages and social-psychological factors on health and environmental risk behaviors, with a focus on culture-based differences and similarities. To this end, Dr. Lapinski has conducted collaborative research projects with her students and colleagues in a number of countries in Asia, the Pacific Rim, Central America, and Africa. Additionally, she has presented work at national and international conferences and published literature in public health and communication journals. Her work has been funded by the National Science Foundation, World Health Organization, and National Institutes of Health.

Matthew T. Lee, Ph.D., is Professor and Chair of Sociology as well as Conflict Management Fellow at the University of Akron. He is vice president of the Institute for Research on Unlimited Love, and he served as vice president of the Center for Restorative Justice of North Central Ohio. He is chair-elect of the Altruism, Morality, and Social Solidarity Section of the American Sociological Association. His research has addressed such diverse topics as altruism and religious experience, immigration and neighborhood levels of violence, social constructions of corporate crime, and the relationship between political opportunities and anarchism. His latest book (coauthored with Margaret M. Poloma and Stephen G. Post) is *The Heart of Religion: Spiritual Empowerment, Benevolence, and the Experience of God's Love.*

Yueh-Ting Lee (Li Yue-Ting), Ph.D., is Professor of Psychology at the University of Toledo. He received his B.A. in English language and American literature from Central South University (CSU) in China and his M.S. in psychology from Beijing Normal University (BNU) before he immigrated to the United States. He received his doctorate in social psychology from State University of New York (SUNY) at Stony Brook in 1991, and he also completed postdoctoral training and research at the University of Pennsylvania in the early 1990s. Dr. Lee has authored and coauthored approximately 90 refereed journal articles and produced several scholarly books, including *Social Cognition: Understanding Ourselves and Others* (with L. Liu); *Leadership and Management in China: Philosophies, Theories and Practices* (with C. Chen); *The Psychology of Ethnic and Cultural Conflict* (with C. McCauley, F. Moghaddam, and S. Worchel); *Personality and Person Perceptions Across Cultures* (with C. McCauley and J. Draguns); and *Stereotype Accuracy: Toward Appreciating Group Differences* (with L. Jussim and M. McCauley). His research has been funded by various federal and state agencies.

Sangeetha Menon, Ph.D., is Professor and Head of the Consciousness Studies Programme at the National Institute of Advanced Studies, on the Indian Institute of Science campus in Bangalore. She is a philosopher-psychologist working in consciousness studies and Indian psychology. Her published work highlights the importance of understanding the nature of complex experiences in the light of global humanistic values such as well-being and personal growth. Her work focuses on the comparative study of emotions, the nature of agency, and the primacy of experience. Dr. Menon's books include *The Beyond Experience: Consciousness in Bhagavad Gita*; *Consciousness, Experience and Ways of Knowing: Perspectives from Science, Philosophy and the Arts*; *Science and Beyond: Cosmology, Consciousness, and Technology in Indic Traditions*; *Consciousness and Genetics*; and *Dialogues: Philosopher Meets the Seer*. Apart from her academic work, she is an artist who is interested in classical dance, design, and photography. Her website is consciousnsss-hop.com.

Craig T. Palmer, Ph.D., earned his doctorate in cultural anthropology from Arizona State University in 1988. His research focuses on incorporating cultural traditions into evolutionary explanations of human behavior. His publications include coauthored books on religion (*The Supernatural and Natural Selection*), sexual aggression (*The Natural History of Rape*), the ecological collapse of the Canadian cod fishery (*When the Fish are Gone*), and altruism (*Kindness, Kinship and Tradition in Newfoundland/Alberta Migration*). He has performed fieldwork in Canada since the 1980s, but is now studying the ability of stories about selfless acts of sacrifice to inspire altruistic acts in others. This current research examines both the general ability of traditional morally elevating stories to inspire evolutionary advantageous altruism among many generations of co-descendants, and the specific ability of stories about individuals who risked their lives to save Jews during the Holocaust to inspire altruism in people from different religious and national backgrounds.

Stephen G. Post, Ph.D., is Professor of Preventive Medicine and Founding Director of the Center for Medical Humanities, Compassionate Care, and Bioethics at Stony Brook University School of Medicine. He founded the free-standing Institute for Research on Unlimited Love in 2001. He has received the Hope in Healthcare Award, the Pioneer Medal from HealthCare Chaplaincy of New York, the Kama Book Award in Medical Humanities from World Literacy Canada, and the Distinguished Service Award from the U.S. Alzheimer's Association. Dr. Post has written eight scholarly books on altruism, compassionate care, and unlimited (*agape*) love, and he is also the editor of nine other books, including *Altruism & Health: Perspectives from Empirical Research* and *Altruism and Altruistic Love: Science, Philosophy and Religion in Dialogue*. He is lead author of the trade book *Why Good Things Happen to Good People: How to Live a Longer, Happier, Healthier Life by the Simple Act of Giving* and the best-selling author of *The Hidden Gifts of Helping*.

Sandi W. Smith, Ph.D., is Director of the Health and Risk Communication Center and Professor in the Department of Communication at Michigan State University. She is associate editor of *Health Communication*. Her work has been funded by the NSF, HRSA, NCI, NIEHS, and U.S. Department of Education. Dr. Smith's research interests center on the impact of communication on health behavior, such as the prevention and detection of breast cancer, persuading people to enroll on state organ donor registries and to engage in family discussion about organ donation, and encouraging college students to consume alcohol moderately. Her research frequently appears in the *Journal of Health Communication* and *Health Communication* as well as in *Communication Monographs, Human Communication Research,* and the *Journal of Communication*, among others. Her book with Steve Wilson, *New Directions in Interpersonal Communication Research*, recently received the G. R. Miller Outstanding Book Award from the National Communication Association. Dr. Smith has been honored with the Distinguished Faculty Award at Michigan State University and received the B. Aubrey Fisher Mentorship Award from the International Communication Association.

Stacy L. Smith, Ph.D., is Associate Professor of Communication at the University of Southern California. Her research focuses on children's responses to mass media portrayals (television, film, video games) of violence, gender, and hypersexuality. Dr. Smith has written more than 50 journal articles, book chapters, and reports on content patterns and effects of the media on youth. Further, she has received multiple "top paper" awards for her research from the Instructional Developmental Division of the International Communication Association. Her research has been written about in *The New York Times*, *Los Angeles Times*, *The Huffington Post*, *Newsweek*, *The Hollywood Reporter*, *Variety*, *Salon.com*, *The Boston Globe*, and *USA Today*. She also has a coauthored essay in Maria Shriver's book, *A Woman's Nation Changes Everything*.

Harry C. Triandis, Ph.D., is Professor Emeritus of Psychology at the University of Illinois. He was the general editor of the six-volume *Handbook of Cross-Cultural Psychology*. He has served as president of the International Association of Cross-Cultural Psychology, the Interamerican Society of Psychology, and the International Association of Applied Psychology as well as of Divisions 8 and 9 of the American Psychological Association. Dr. Triandis was a Guggenheim Fellow, a Distinguished Fulbright Professor to India, and a Fellow of the American Association for Advancement of Science. He has received the American Psychological Society's James M. Cattell Award, a Lifetime Contributions Award from the Academy of Intercultural Research, and a Lifetime Contributions Award from the Society for Personality and Social Psychology. His 2009 book *Fooling Ourselves* received the William James Award from Division 1 of the American Psychological Association.

About the Editor

Douglas A. Vakoch, Ph.D., is Professor of Clinical Psychology at the California Institute of Integral Studies as well as Director of Interstellar Message Composition at the SETI Institute. He has served on the Executive Committee of the Society for Cross-Cultural Research, which brings together psychologists, anthropologists, and other social scientists to foster interdisciplinary dialogue and research. Dr. Vakoch chaired the symposium "Altruism in Cross-Cultural Perspective" at the 34th Annual Conference of the Society for Cross-Cultural Research, which gathered researchers from anthropology, psychology, sociology, medicine, and public health for the first in-depth discussion of altruism across cultures. He has been a faculty participant in two separate month-long summer seminars sponsored by the John Templeton Foundation: "Biology and Purpose: Altruism, Morality, and Human Nature in Evolutionary Theory" and "Works of Love: Scientific and Theological Perspectives on Altruism." Dr. Vakoch is licensed as a clinical psychologist, and he serves on the editorial board of the journal *Ecopsychology*. His research spans the fields of psychology, anthropology, environmental studies, and space sciences, and his books include *Psychology of Space Exploration: Contemporary Research in Historical Perspective* (NASA, 2011); *Communication with Extraterrestrial Intelligence* (SUNY Press, 2011); *Ecofeminism and Rhetoric: Critical Perspectives on Sex, Technology, and Discourse* (Berghahn Books, 2011); *Feminist Ecocriticism: Environment, Women, and Literature* (Lexington Books, 2012); *On Orbit and Beyond: Psychological Perspectives on Human Spaceflight* (Springer, 2013); *Archaeology, Anthropology, and Interstellar Communication* (NASA, 2013); *Astrobiology, History, and Society: Life Beyond Earth and the Impact of Discovery* (Springer, 2013); *Extraterrestrial Altruism: Evolution and Ethics in the Cosmos* (Springer, 2013).

D.A. Vakoch (ed.), *Altruism in Cross-Cultural Perspective*,
International and Cultural Psychology, DOI 10.1007/978-1-4614-6952-0,

Index

D.A. Vakoch (ed.), *Altruism in Cross-Cultural Perspective*,
International and Cultural Psychology, DOI 10.1007/978-1-4614-6952-0,

CPSIA information can be obtained at www.ICGtesting.com
Printed in the USA
LVOW070509010613

336489LV00007B/70/P

9 781461 469513